A Personal History of
CESR and **CLEO**

The Cornell Electron Storage Ring and
Its Main Particle Detector Facility

A Personal History of
CESR and CLEO

The Cornell Electron Storage Ring and
Its Main Particle Detector Facility

Karl Berkelman

Cornell University, USA

World Scientific

NEW JERSEY • LONDON • SINGAPORE • SHANGHAI • HONG KONG • TAIPEI • BANGALORE

Published by

World Scientific Publishing Co. Pte. Ltd.

5 Toh Tuck Link, Singapore 596224

USA office: Suite 202, 1060 Main Street, River Edge, NJ 07661

UK office: 57 Shelton Street, Covent Garden, London WC2H 9HE

British Library Cataloguing-in-Publication Data
A catalogue record for this book is available from the British Library.

A PERSONAL HISTORY OF CESR AND CLEO

ISBN 981-238-697-1

This book is printed on acid-free paper.

Printed in Singapore by Mainland Press

Preface

———————————— • ————————————

I have attempted in this book to present a historical account of the Cornell Electron Storage Ring and its main detector facility CLEO from their beginnings in the late 1970s until the end of data collection at particle energies above the threshold for B meson production in June 2001. The CESR electron–positron collider was the culmination of a series of electron accelerators constructed at the Cornell Laboratory of Nuclear Studies starting in 1945. The measurements made on the products of the e^+e^- collisions were performed with the multipurpose CLEO apparatus, built and operated by the CLEO collaboration, consisting of about 200 faculty, staff and graduate students from over 20 universities.

My account is based mainly on my recollections as a participant and on documents readily available to me. It is therefore unavoidably biased, and probably emphasizes unduly events in which I was personally involved. I have tried to be systematic and as complete as possible, given the constraints of length, but there are surely omissions, inaccuracies, and lapses in my memory. An earlier version was written mainly for physicists, particularly for new members of the Cornell Laboratory or the CLEO collaboration who may be curious about how we got to where we are. In this book version I have tried to broaden the accessibility to a wider audience. Frequently used terms are collected in a glossary. Much of the detailed information, only of interest to particle physicists, is presented in the appendix.

I will start in the first chapter with a quick and rather superficial review of the major events in particle and accelerator physics in the decades preceding the conception of CESR and CLEO. Readers who would like to learn more about the players and events of 1930–1980 period can consult the earliest of the listed references. Particle physicists can skip to the second chapter.

Contents

The Rise of Accelerators

--- • ---

Accelerators are machines that use electromagnetic forces to increase the energy of charged particles. In modern machines, the velocities of the particles can approach very close to the velocity of light, that special relativity tells us is the maximum possible. To be clear in this limiting case on what we mean as acceleration, we need to be familiar with the relativistic relations among velocity, energy, momentum and mass, as well as the units that are commonly used to express them quantitatively.

Velocities are conveniently given in terms of the fraction of the velocity of light:

$$\beta = v/c, \quad \text{where } c = 3 \times 10^8 \text{ m/sec}.$$

A particle of fixed mass m has rest energy mc^2. If it is moving, it also has kinetic energy which depends on its velocity. The total energy of a particle, rest energy plus kinetic energy, is

$$E = \gamma mc^2 = \frac{1}{\sqrt{1 - \beta^2}} mc^2.$$

Notice that an accelerator can increase the energy of a particle without limit as β approaches one, even though the velocity hardly increases. The particle is not actually "accelerating" much in the everyday sense of the word, even when its energy is increasing significantly. It would take an infinite amount of energy to bring the velocity up to $\beta = 1$.

Momentum, which is zero for a particle at rest, increases without limit as the particle is accelerated:

$$P = \gamma \beta mc = \frac{\beta}{\sqrt{1 - \beta^2}} mc.$$

If the force comes from an electric field $\vec{\mathcal{E}}$, the time rate of increase of momentum is $d\vec{P}/dt = q\vec{\mathcal{E}}$, where q is the charge of the particle. The corresponding energy increase can be expressed in terms of the voltage difference across which it travels: $\Delta E = q\Delta V$.

We find it convenient to measure all energies in terms of the voltage that would be required when the particle charge has the magnitude of the electron charge e. This energy unit is called the electron volt. One usually encounters it in multiples: keV, MeV, GeV for 10^3, 10^6, 10^9 eV. Rest energies mc^2 of particles are also measured in these same units — 0.511 MeV for the electron, 938.272 MeV for the proton — and we call it "mass" rather than rest energy. We further simplify by measuring momentum$\times c$ in the same units and calling it "momentum". When measured in these units, the relation $E^2 - (Pc)^2 = (mc^2)^2$ becomes simply $E^2 - P^2 = m^2$, the particle velocity is $\beta = P/E$, and the Lorentz factor γ becomes E/m.

One of the early motivations for accelerating particles was to improve on Rutherford's pioneering experiment, that is, to study the spatial distribution of charge or nuclear matter by scattering a beam of particles from various nuclear targets. The principle is related to that of the electron microscope, although one has to reconstruct the image mathematically rather than optically. The spatial resolution is comparable to the wavelength λ, which for a particle of momentum p is h/p (h is Planck's constant). The higher the beam energy or momentum, the finer the resolution. In other studies of nuclear structure the goal was to break up an atomic nucleus, for instance, by adding a proton to form an unstable compound nucleus. To accomplish this, the proton has to have enough energy to overcome the electrostatic repulsion of the target nucleus. In recent decades an important goal has been to study new forms of matter by using collisions of high energy particles to convert kinetic energy to mass (rest energy) in the creation of new particles: mesons, hyperons, and such. With higher beam energy, one can make more massive particles.

So for these reasons and more, the most important parameter characterizing the capability of an accelerator is not the maximum velocity it can give to a particle but the maximum energy. At the time of writing, about 1000 GeV is held by the Fermilab Tevatron proton accelerator. Almost as important is the beam intensity, the number of particles that can be accelerated per unit of time, for instance up to 10^{14} per second for the Fermilab machine. If the intensity is insufficient, you cannot detect rare processes, and measured reaction rates may be dominated by the random fluctuations of small numbers.

We should also include as a parameter the kind of particle the machine is designed to accelerate. In principle, it could be any charged particle that lasts long enough to survive the acceleration. In elementary particle physics, this means the proton or electron, and in recent years their antiparticles, the antiproton and the positron. For nuclear studies, though, we can accelerate practically any kind of ion. Heavy ions from deuterium to gold have been accelerated.

The first accelerators [1] used a DC voltage in a vaccum to accelerate protons across the gap from one electrode to the other. The maximum particle energy was limited to a few hundered keV (a few MeV in later years) by breakdown of the voltage. It was E.O. Lawrence in the 1930s who conceived the idea of circulating protons or other positive ions in a magnetic field so they could make multiple passes through a more modest electric field, gaining energy on each pass. The electric field would oscillate and the particle bunch passage through the voltage gap would be synchronized to occur at the right polarity. As the proton velocity increased, it maintained synchrony with the oscillating electric field, because the increasing momentum caused the orbit radius in the magnetic field to increase.

Lawrence's cyclotron at UC Berkeley inaugurated the era of circular accelerators and eventually a new style of experimental physics. The energy of the early cyclotrons, however, was limited to several tens of MeV. As protons become relativistic the simple proportionality of momentum and velocity breaks down and the arrival of the particle bunch at the gap gets out of phase with the voltage oscillation. But the demand from the experimenters (who in those early days were also the accelerator builders) was always for more energy. The solution was to modulate the frequency of the accelerating voltage to keep step with an accelerating bunch of protons. In the 1940s and 1950s synchrocyclotrons were built at a number of universities in the US, at at the CERN international laboratory in Geneva, Switzerland, and at the Dubna laboratory in Russia. As the energies reached several hundreds of MeV, the size of the magnet needed to accommodate the particle orbit at the highest momentum got to be a very significant cost factor, again limiting the maximum energy achievable.

This time the solution was to let the magnetic field vary in time in proportion to the particle momentum, keeping the orbit radius fixed, instead of letting the orbit spiral to larger and larger radii in a fixed magnetic field. Thanks to the principle of phase stability discovered by McMillan and Veksler in 1945, the particles would always gain momentum at the accelerating voltage gap at just the right rate to match the rate of increase in magnetic field. After the accelerator had produced a beam pulse at the peak energy, the magnetic field would return to its low starting value and the cycle would repeat with a new bunch of particles. For large energies a ring shaped magnet to accommodate the fixed radius synchrotron orbit was much more economical than providing the magnetic field over the full area enclosed by the maximum radius in a synchrocyclotron. The first synchrotrons accelerated electrons, and such machines were built with energies of hundreds of MeV at several universities, including Cornell. Proton synchrotrons of several GeV energy were built starting in the late 1950s. The cost had become too high, though, for a single university, so some, such as UC Berkeley, became national laboratories and others formed consortia to run new national laboratories — Brookhaven on Long Island and later Fermilab in Illinois, with funding from the Department of Energy, formerly ERDA, and before that, the AEC. By the 1970s the only remaining frontier-energy accelerator not at a DOE national laboratory was the 10 GeV electron synchrotron on the Cornell campus, supported by the National Science Foundation.

Meanwhile, outside the US, proton synchrotrons came into operation at CERN (on the Swiss–French border), Saclay (France), Rutherford Laboratory (England), Dubna and Serpukhov (Russia), and KEK (Japan). Electron synchrotrons were running at Frascati (Italy), Bonn and Hamburg (Germany), Daresbury (England), and Lund (Sweden).

2

Synchrotrons and More Synchrotrons, up to 1975

——————————————— • ———————————————

In 1946 several Cornell faculty returned from their wartime work with the Manhattan Project at Los Alamos and launched the Laboratory of Nuclear Studies (LNS) as a research subunit of the Physics Department. Among them were experimentalists Robert Bacher, Boyce McDaniel, Dale Corson, John DeWire and Bill Woodward and theorist Hans Bethe. There had already been a nuclear physics research program at Cornell before the war, started by E.O. Lawrence's former student, Stanley Livingston [1]. The experimental program centered on the first cyclotron built outside of Berkeley, a 1 MeV machine in the basement of the physics building, Rockefeller Hall (see Fig. 1). There is a story that when John D. endowed the building around 1900, he had intended to donate more money later for laboratory equipment. This was not realized at Cornell, so the university scrimped on the building in order to have enough left over to equip it. When Rockefeller saw what had been built, he was disappointed and never gave Cornell another cent.

The new Laboratory needed a new building. Although the money to pay for it was not yet in sight, the Cornell president, Edmund Ezra Day gave the go-ahead for construction, presumably to keep these hot-shot physicists from leaving Cornell. By the time the building was completed he found a wealthy alumnus, Floyd R. Newman, to pay for what is now Newman Laboratory.

The founders of the LNS had worked at Los Alamos with a larger cyclotron that had been set up there under the leadership of Robert Wilson, another ex-student of E.O. Lawrence. It was natural then to plan for a higher energy successor to the prewar Cornell cyclotron. But instead of building another proton accelerator, they opted for an electron synchrotron. I do not know what the motivation was or who originated the idea, but it was a fateful decision as seen from our present point of view a half century later. Maybe it was the hope of getting more bang

5

Fig. 1. Boyce McDaniel (middle) and Robert Wilson (right) donating the Cornell 1 MeV cyclotron to Prof. Solly Cohen of the University of Jerusalem, in 1954.

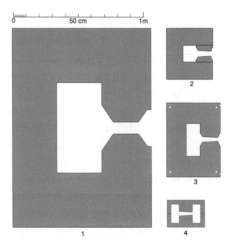

Fig. 2. The guide-field magnet cross-section, to scale, for each of the Cornell synchrotrons: the 300 MeV (1), 1 GeV (2), 2.2 GeV (3), and 10 GeV (4).

(energy) for the buck. When the first LNS director, Robert Bacher, left Cornell almost immediately to become a member of the newly formed US Atomic Energy Commission, it was natural to replace him by hiring Robert Wilson away from Harvard — another fateful decision. Wilson was known from the Los Alamos days for his ambition, creativity and can-do spirit.

The electron synchrotron was funded by the Office of Naval Research and was completed in 1949 with a maximum energy of 300 MeV. It had an orbit radius of

1 meter, but actually looked much larger, because the magnetic field on the orbit was provided by bulky **C** shaped magnet sections (Fig. 2) with the gap for the donut shaped beam tube facing the center of the machine.

In those days, thanks to Yukawa, the meson was seen as the carrier of the strong nuclear force binding protons and neutrons in nuclei in much the same way that the photon is the fundamental carrier of the electromagnetic force binding electrons in an atom. The study of meson interactions was the highest priority for the understanding of nuclear forces. Athough at first physicists did not yet know the difference between the strongly interacting pi meson and the muon seen in cosmic rays, they knew that they would need several hundred MeV of electron energy to create mesons. At an electron accelerator the idea was first to make a beam of photons through the bremsstrahlung process by running the electrons into a high-Z material, then to use the high energy end of the photon (γ) spectrum to photoproduce mesons (π) from a nuclear target (N): $\gamma + N \rightarrow \pi + N'$. The yield of mesons was found to be small but increasing rapidly with beam energy. This and the fact that there were by then competing electron synchrotrons in the same energy range at Berkeley, Caltech, MIT, Michigan and Purdue inspired Bob Wilson to start thinking about how to make a major jump in beam energy. He was determined that Cornell would be on the frontier. He had to beat the competition.

To increase the energy from 300 MeV to say 1 GeV would mean increasing the orbit circumference from 6 meters to 28 meters. Without some new idea, that would mean just that much more magnet mass and cost. Wilson's idea was to decrease the transverse size of the donut shaped vacuum chamber for the beam and thereby reduce the gap height and width of the magnet (see Fig. 2). He managed to come up with a low enough magnet mass and power requirement to allow use of the existing power supply, provided he could come up with a suitable choke. For the choke he proposed to use the magnet of the 300 GeV machine. Nothing would be wasted! Cutting down the beam aperture so drastically was certainly a risky plan, since the beam spreads out transversely and needs room to survive. Wilson had the foresight, however, to design the magnet with removable pole pieces, so that the gap width and height could be tailored later when one would know just how much room the beam actually needed. Of course, if the donut cross-section had to be larger than the original plan, the magnet would have to run at a lower excitation to avoid saturation, and the maximum beam energy would be less. Wilson was never one to shy away from risk. Anyway, the cost saving was enough to allow the ONR to fund it, and the 1 GeV synchrotron started construction.

At about this time, Nicholas Christofilos in Greece came up with the idea of alternating gradient focusing. In this scheme the magnetic guide field in a synchrotron varies spatially in such a way as to provide a strong focusing force on particle trajectories, keeping them much closer to the ideal orbit in the accelerator. Thus the magnet gap height and width could be made much smaller. What a lucky break! Before the synchrotron was completed, Wilson redesigned the magnet pole pieces for alternating gradient and smaller aperture, and the Cornell 1 GeV synchrotron started operating in 1954 as the first alternating gradient synchrotron. In 1955 I arrived at Cornell as a graduate student and started my thesis work on this synchrotron under Prof. Dale Corson, looking at neutral pion photoproduction. For my highest energy data point, I was able to push the synchrotron energy almost to 1.4 GeV. As I was finishing my degree Dale Corson left to become Dean of Engineering and then from 1969 to 1977 President of the University.

I remember one occasion while I was a graduate student working at the 1 GeV machine when the air-cooled coils of the 300 MeV magnet (now serving as a choke) overheated and the insulation caught fire. McDaniel and I were the only people in the lab that night. Mac climbed down into the hole in the middle of the magnet ring while I ran around the lab gathering up all the CO_2 fire extinguishers I could find and handing them down to him. We succeeded in putting the fire out.

Initially the work at the 1 GeV synchroton concentrated on meson photoproduction, first single pion (π^+ and π^0) photoproduction, then photoproduction of heavier mesons η and K^+ and multiple pions. The urge to push the energy beyond 1 GeV was a result of the competition from the rebuilt Caltech synchrotron and the 1 GeV Stanford linear accelerator. Robert Hofstadter at Stanford was in the midst of his Nobel Prize investigation of the charge and magnetic moment distribution of the proton through the elastic scattering of electrons from hydrogen. Hofstadter's claim that this could only be done with a linear accelerator and that a circular machine was good only for studying photoproduction with bremsstrahlung beams was a challenge to Bob Wilson. So he and I and Jim Cassels, on leave from Liverpool, England, set out to measure the proton form factors at higher energies than those accessible at the Stanford linac. The biggest problem with using the electron beam directly is getting it out of its circular orbit to where it can hit a target. It was much easier to put a bremsstrahlung target inside the beam chamber and deflect the beam slightly to hit the target (or vice versa) at the end of the acceleration cycle. But for electron–proton scattering we needed a hydrogen target. Rather than trying to extract the electron beam from the ring to run through an external liquid hydrogen target, we used a toothpick sized piece of polyethylene (essentially CH_2) inside the beam chamber. We set up magnetic spectrometers for electron and the recoil proton emerging from the target and used them in coincidence to constrain the kinematics of the reaction and reject scatters from the carbon in the target. We observed the forward bremsstrahlung from the target to monitor the effective incident electron flux, which was enhanced by multiple passes. Eventually we succeeded in measuring the form factors at record values of momentum transfer. Although the target was set up in a field-free straight section between synchrotron magnets, some of our early data were falsified by the fact that the magnetic fringe fields distorted the calculated angles between scattered electron and recoil proton — that was my fault. Later I improved the experiment by designing a liquid hydrogen target that could be inserted into the beam vacuum chamber.

Although the 1 GeV synchrotron was successful in reaching and surpassing its energy goal, we were frustrated by its unreliability. As Wilson said [2], "If we had any secret in constructing machines rapidly and at not great cost, that secret was our willingness, almost our eagerness, to make mistakes — to get a piece of equipment together first and then to change it so that it will work. ... something that works right away is over-designed and consequently will have taken too long to build and have cost too much." The rubber-gasketed glass vacuum chamber was frequently leaking and occasionally imploding. Keeping the magnet pole tips aligned was a challenge. At high energies the skimpy magnet (Fig. 2) limited the good field region. The injection energy provided by the catalog-ordered 2 MeV Van de Graaff accelerator was too low for stable beam behavior in the synchrotron ring. Although the latter problem was solved by buying a 10 MeV linac from ARCO, we made plans to rebuild the synchrotron with a new well-engineered magnet (Fig. 2) and vacuum chamber. In an effort to keep ahead of the competition, we excavated more space and enlarged the ring to accommodate a 2.2 GeV ring.

Somehow the money to pay for this was scraped together out of normal operating funds, and we mobilized to get the work done. McDaniel was in charge. I remember one day when professors, students and technicians all pitched in to dismantle the old ring — by the end of the day it was all out the door and the floor was clear. In three months we marked the completion of the new ring with a group photograph (Fig. 3). On an evening in April 1964 we were ready to try for beam. As soon as

Fig. 3. Group portrait on the completion of the Cornell 2.2 GeV synchrotron ring in 1964. From left to right are Don Edwards, Al Silverman, Jack Kenemuth behind Bill Woodward, Maury Tigner behind Boyce McDaniel, Raphael Littauer behind Bob Wilson, Bob Anderson, Bruno Borgia, Peter Stein, Erwin Gabathuler, Karl Berkelman and John DeWire.

the magnet was excited and the injection timing adjusted, a good beam appeared! Wow! We could hardly believe it. We trooped off to the nearest bar and spent the rest of the evening celebrating.

While we were building the 2.2 GeV synchrotron and exploiting it for physics (more photoproduction and electron scattering, including the measurement of the charge form factor of the pion), Bob Wilson was designing the next synchrotron. We had to regain the frontier, now that the 6 GeV Harvard–MIT Cambridge Electron Accelerator (CEA) was being completed on the Harvard campus. The goal was 10 GeV. Again, the only hope of getting it approved by the National Science Foundation, which had taken over support of the Laboratory program from ONR, was keeping the cost low. Since the magnet ring circumference obviously had to be scaled way up, Wilson had to play the old trick of reducing its cross-sectional area. His new idea was to make an H magnet (Fig. 2) instead of a C magnet — that is, with flux returns on both sides of the gap — and to put the whole magnet — iron and coils — *inside* the beam vacuum chamber. Without chamber walls inside, the gap could be made smaller and the magnet cross-section could be reduced without sacrificing space for the beam. Also, he further decreased the magnet cross-section by lowering the maximum field value. Although this necessitated increasing further the ring circumference, the cost penalty was offset by savings in the rf (radio-frequency) accelerating system required to provide the reduced energy radiated. Figure 2 shows the comparison of the 10 GeV magnet with previous Cornell synchrotron magnets.

The basement of Newman Laboratory would no longer be adequate to house the half-mile circumference. It seemed that the site would have to be off-campus. Wilson, however, was adamant in insisting that the new ring should be close to Newman Lab and the rest of the Physics Department. Eventually he succeeded in getting approval from the university administration to tunnel 50 feet under the intramural athletic fields (Fig. 4). And somehow he convinced the NSF to provide (in April 1965) $12 million for the new 10 GeV synchrotron, linac injector, tunnel and laboratory building. This was quite a feat in those days. I would like to know how he did it. It was also an important milestone. Without this approval the later CESR and CLEO would never have existed.

A tunnel was a novelty for high energy accelerators. Previous rings had been located in cut-and-cover excavations. It was a crucial item in the cost estimate, too, and there was enormous relief when the winning $1 million bid came in after several others in the $1.8–2.0 million range. As Wilson says [2], "The rest of the bids were anticlimatic and after they had been opened, Mr. Traylor stood up, the obvious winner. His face was wreathed in joy, but tears were literally streaming from his eyes. The next lowest bid was $1.5 million and he sadly explained that he had just dropped half a million on the table." Actually, he had just finished a big sewer for the city of St. Louis and he really needed something to keep his tunnel boring crew occupied before the next big sewer contract. They did a beautiful job

Fig. 4.   1994 aerial view of Cornell University and Cayuga Lake, looking NNW. The oval shows the location of the tunnel for the 10 GeV synchrotron and the Cornell Electron Storage Ring. The building at the south side of the ring is Wilson Laboratory; Newman Lab is at the left edge of the picture, just in front of the seven-story chemistry research building.

and were quite proud of the opportunity to dig something classier than a sewer. One of their business personnel, Mr. Tom Allen, took an administrative job with us, and until his retirement in 2000, he was the financial watchdog for the Lab.

Just as the tunneling was finished, Wilson left Cornell to become the director of the new National Accelerator Laboratory (later renamed Fermilab) in an empty field in Illinois. He had criticised the earlier Berkeley proposal for a new proton synchrotron as offering too low an energy at too high a price. So when he was challenged to do better, he had to accept. Boyce McDaniel, the Associate Director in charge of the construction of our new machine, took over as Director of LNS. In many ways Wilson and McDaniel were opposite. Bob was the idea man, the charismatic leader, willing to take huge risks to be at the forefront of high energy physics. He was at his best in setting the goals, providing the grand design, and inspiring his colleagues (including the funding agencies and their advisors) to join in the struggle to meet the goals. He sometimes made mistakes, but with the help of Mac and the rest of the Laboratory team, he always overcame them. Mac was a careful, systematic problem solver, a wizard at designing state-of-the-art accelerator systems and making them work, and he was a superb team manager. I do not think Mac ever made a mistake. As different as they were, Bob and Mac each had a profound respect for the other. Several years later, when Bob ran into trouble with some of his riskier ideas in the construction of the Fermilab accelerator, he called on Mac to come to Illinois for a while and set things right there.

One of the key components of any circular accelerator is the setup for providing the push for the particles, that is, the radio-frequency (rf) system. Since circulating electrons radiate energy with a flux that varies as the fourth power of their energy, replacing that energy gets to be a major cost factor the higher the energy of a synchrotron. Neither Wilson nor McDaniel were expert at rf, but fortunately Maury Tigner was. Maury had been Wilson's graduate student, and although Maury and I worked together briefly on a electron scattering experiment after he got his degree, he soon specialized in accelerator technology, and specifically rf accelerating systems. Maury was the first to adapt the linac concept, that is, a multiple-pass traveling wave array of coupled rf cavities, to provide the accelerating mechanism for a synchrotron. He applied it first to the 2.2 GeV machine, then to the new 10 GeV machine. Maury called the system a "synac", but the name did not stick.

Cornell's fourth synchrotron, the "Ten GeV Machine" was completed in 1967 (see Fig. 5) and worked well [4]. The energy even reached 12 GeV eventually. The Laboratory of Nuclear Studies had managed to keep itself at or near the forefront of particle physics [3] for the two decades since its founding by periodically rebuilding its accelerator facilities (Fig. 6). The early Ten-GeV experimental program was carried out mainly by physicists from Cornell, Harvard and Rochester, and included a wide-angle bremsstrahlung test of quantum electrodynamics (QED) using the internal electron beam impinging on a target in the beam chamber, and a number of meson photoproduction experiments using the external bremsstrahlung beam: wide-angle e^+e^- and $\mu^+\mu^-$ tests of QED and production of π^+, ρ^0, ω, ϕ, and ψ mesons.

Once QED had passed several high energy tests, and the photoproduction cross-sections of the low-lying mesons had been measured, we concentrated attention on meson electroproduction using an extracted electron beam. During this period our chief competition was from the Stanford 20 GeV electron linear accelerator, which had opened up the field of deep inelastic electron–nucleon scattering. We were motivated by the desire to see what the nucleon fragments looked like in the rather

Fig. 5. Hans Bethe and Boyce McDaniel bicycling around the newly completed 10 GeV synchrotron ring.

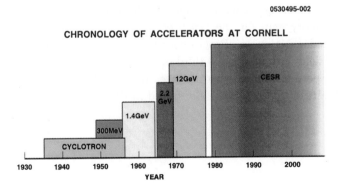

Fig. 6. Chronology of accelerators at Cornell.

copious yield of pointlike electron–parton collisons. The electrons in the lower energy Cornell machine did not really have short enough wavelengths to resolve the constituents of the nucleon and explore the deep inelastic kinematic range, though. The electroproduction cross-sections were instead dominated by virtual-photon-plus-nucleon energies in the nucleon resonance region, and the interaction of the photon with the target was telling us more about the vector meson nature of the photon than about the more interesting pointlike constituents of the nucleon. This prompted us to upgrade the beam energy from 10 GeV to 12 GeV by adding more radio-frequency accelerating cavities, but 12 GeV was still much smaller than the 20 GeV available at SLAC.

An advantage we could claim over Stanford was the fact that coincidence experiments, such as required to see the nucleon fragmentation products along with the scattered electron, were much easier with the few percent beam duty cycle of a synchrotron than with the 10^{-5} duty cycle of the Stanford linac. It is easier to tell if two particles are produced simultaneously when the collisions are spread out over time than when they are occurring in short pulses. The SLAC physicists, however, were getting clever at overcoming this obstacle and Martin Perl's group had actually performed a successful multiparticle coincidence electroproduction experiment. Moreover, the physics of the hadronic final states in deep inelastic scattering was turning out to be rather uninteresting. Once a parton had been punched out of a nucleon, it fragmented into a hadron jet in a way that depended mainly on the total energy available, with little or no memory of how it was produced. Multiplicities, for example, were insensitive to the q^2 of the virtual photon. High energy electroproduction final states looked just like the debris of any high energy hadronic collision.

So we began to look for something better to do. Lou Hand went off to do deep inelastic muon–nucleon scattering at much higher energies at Fermilab. Bernie Gittelman (in 1975–1976) and I (in 1974–1975) spent our sabbatical years at the DORIS e^+e^- storage ring at DESY. Maury Tigner started up a research group to develop superconducting rf cavities, which would be the only plausible way to make a significant energy gain for the synchrotron. The proposal for 1974–1975 NSF funding included a section on "Program for Energy Increase" which mentioned that "A new guide field with about one third of the perimeter dedicated to accelerating cavities might permit operation to the level of about 25 GeV." A superconducting rf cavity had already been successfully tested in the synchrotron in 1974.

3

The CESR Proposal, 1974–1977

The idea of building a storage ring in which two oppositely directed beams would circulate and produce beam–beam collisions had been in the air at Cornell ever since Gerry O'Neill first suggested it in the 1950s. One could achieve higher center-of-mass collision energies $E_{cm} = E_1 + E_2$ in a beam–beam collision, than in the collision of a beam with a fixed target where $E_{cm} = \sqrt{2(E_{\text{beam}} + M_{\text{target}})M_{\text{target}}}$. This was especially true for the case of an electron target, where $M_{\text{target}} \ll E_{\text{beam}}$. Bob Wilson had assigned to Maury Tigner in 1959 the task of building a table-top electron storage ring (Fig. 7) as a PhD thesis topic in accelerator physics. This was at the same time that Bernie Gittelman was participating in the operation of the first e^-e^- storage rings at Stanford, and the first e^+e^- rings were being built at Frascati and Novosibirsk. Although Tigner had made a preliminary conceptual design for a Cornell e^+e^- storage ring in 1973 (see Fig. 8), there was still a lot of skepticism at Cornell. It is difficult to store beams that are intense enough to provide a usable rate of collisions. The beams do not remain stored forever; the intensities are depleted not by the desirable beam–beam collisions but by collisions with residual gas atoms in the beam chamber. It was not clear that one could do more than just measure the total annihilation cross-section, which was expected to be getting smaller as the inverse square of the beam energy. Hardly anyone expected that there would be any useful physics beyond checking quantum electrodynamics with processes like bremsstrahlung, e^+e^- pair production, annihilation and scattering. QED had passed many accurate experimental tests, so most physicists could not get excited about the prospect of more tests at higher energies.

However, as time went on and storage ring data came in from the Adone ring at Frascati, from the CEA Bypass, and eventually from the SPEAR collider at SLAC, some of us at Cornell became convinced that the future of the lab lay in

Fig. 7.   Three quadrants of the storage ring that Maury Tigner built for his PhD project.

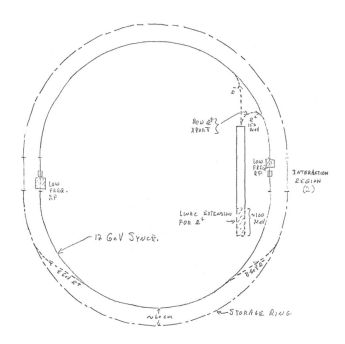

Fig. 8.   The first drawing of CESR, from "A possible e^+e^- storage ring for the Cornell synchrotron", by Maury Tigner, April 1973.

building a storage ring in the 10 GeV tunnel, using the synchrotron as an injector. In fact, the CEA Bypass data on the total e^+e^- cross-section, published in 1973, surprised everyone by showing a rise with increasing energy instead of the expected $1/E^2$ dependence. I recall Bjorken's talk at the Bonn conference in August 1973, in which he speculated that a hypothetical fourth quark was being produced.

Interest in storage rings was considerably reinforced by the November Revolution, that is, the 1974 discovery of the J at the Brookhaven AGS proton synchrotron by Sam Ting's group, and the discovery of the ψ at the SPEAR e^+e^- collider by Burt Richter and collaborators. It was actually the same particle, now officially known as the J/ψ to assuage both of the discoverers' egos. The name ψ seems to follow naturally from the analogous ϕ meson at lower mass, and I am told that the Chinese character for Ting looks like J. To avoid the ugly J/ψ notation, I usually use the J name when it is produced in hadron collisions and ψ when it is produced by e^+e^-.

The sight of that colossal resonance at 3.1 GeV e^+e^- energy convinced the doubters here at Cornell, including cautious Boyce McDaniel, that there was exciting physics in electron–positron collisions. It set us on the course to building CESR. The most popular physical interpretation of the discovery was that the CEA had earlier seen the threshold for a new "charmed" quark and that the ψ was the resulting $c\bar{c}$ bound state.

The energy of a Cornell ring would be determined by the circumference of the existing tunnel. The economics of the rf power requirements dictated that it would have to be somewhat lower than the synchrotron energy, say 8 GeV per beam. Although the question as to whether there might be useful physics for a Cornell collider seemed to be settled, at least to our satisfaction, the question of whether one could inject enough positrons from a synchrotron remained. SPEAR at SLAC, the most successful storage ring, circulated a single bunch of electrons and a single bunch of positrons. Positrons were produced in a showering target part way down the two-mile SLAC linac and then accelerated to the SPEAR energy in the reversed phased remainder of the linac. Since the linac injector for the Cornell synchrotron had only 150 MeV total energy, a target part way along its length would produce a relatively meager flux of positrons, and it would take much too long to build up a single intense bunch in the storage ring by repetition of the sequence of single bunch positron production, acceleration in the linac, acceleration in the synchrotron, and injection into the storage ring.

Maury Tigner came to the rescue with a fiendishly clever "vernier coalescing" scheme (Fig. 9). Although the Cornell linac could make only a rather low number of positrons in a single bunch, it would take no longer to fill the storage ring with about 60 such bunches, equally spaced around the ring Suppose the storage ring was designed to have 61/60 times the circumference of the synchrotron. You could then extract bunch #2 from the storage ring, send it back to the synchrotron for one time around, inject it again into the storage ring and it would fall on top of bunch

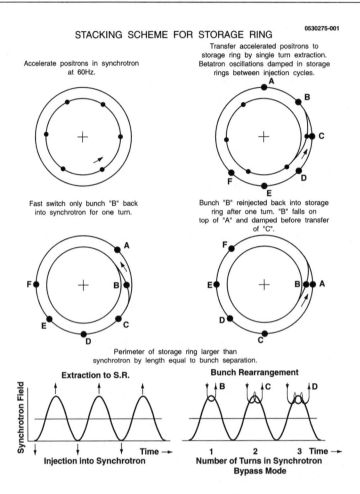

Fig. 9. Diagram explaining the vernier coalescing scheme for positron injection into CESR, from "Improved method for filling an electron storage from a synchrotron", by Maury Tigner, CLNS-299, February 1975.

#1, which had been 1/61 of the circumference ahead of it. Then bunch #3 would be diverted through the synchrotron for two circuits, again falling on top of bunch #1 in the storage ring, and so on until all of the 60 bunches were coalesced into one intense bunch. The whole coalescing procedure could be done in a few seconds. It required very fast pulsed magnets to accomplish the ejection and injection with the whole sequence under precise computer control, but there was no reason why it could not be done. This was what we needed to convince ouselves and the rest of the physics community that we had a practical plan for achieving the required beam currents.

One bunch of electrons and one bunch of positrons circulating in opposite directions along the same path will collide at two diametrically opposite points. By correct phasing of the bunches we could arrange one of the two intersection points to occur in the large L-0 ("L-zero") experimental hall on the south side of the ring. The other would occur in the much smaller L-3 area in the north. In the tunnel, the new ring would be on the outside wall, opposite the synchrotron, and their beam lines would be 1.5 m apart typically (Fig. 13, top). There was room in L-0 to make a bulge in the storage ring layout to bring the intersection point far enough away from the synchrotron to accommodate a large detector, but in the north the two rings would be no further apart than they were in the tunnel.

Six months after the announcement of the discovery of ψ, in May 1975 the Lab submitted "A Proposal to the National Science Foundation for Construction Funds to Modify the Cornell Electron Synchrotron Facility to provide an Electron–Positron Colliding Beam Capability". The text of the document summarized the CESR design parameters, magnet, vacuum system, rf system, injection and controls [6]. The luminosity goal was 10^{32} cm^{-2} sec^{-1} at 8 GeV per beam, the same as for the higher energy PETRA and PEP rings proposed at that time. The total project cost, estimated at \$16.8 million, did not include detectors, but it was stated that "Very sizeable capital investment and annual operating costs will be required to provide the experimental equipment and support the staff of the various experimental programs." The stated physics goals included heavy quarks and leptons, spectroscopy of hadronic resonances, hadronic fragmentation, electroweak effects in annihilation processes like $e^+e^- \rightarrow \mu^+\mu^-$, photon–photon collisions, and high energy tests of QED. Also discussed were prospects for a synchrotron radiation facility. The total cost included \$1.1 million in civil construction to enlarge the north experimental area.

The problem would be selling the idea to the National Science Foundation. We had several arguments for such an $8 + 8$ GeV e^+e^- storage ring.

- It would extend the usefulness of the Cornell synchrotron for less than \$20 million.
- It would fill the energy gap between the 2.5 to 5 GeV per beam SPEAR and DORIS rings and the PEP and PETRA rings at 14 to 18 GeV beam energies proposed by the SLAC and DESY labs. It was known already that a beam-beam collider ring (in contrast to the beam-on-fixed-target accelerator) operates with optimal beam luminosity only when the energy is set close to its maximum.
- It would provide opportunities to test the predictions of quantum electrodynamics further in electron–positron collisions and to look for new phenomena.

Although we were convinced that it would be good for the Cornell Lab, I thought at the time that the physics case was rather weak. Once the ψ had been found and the list of known quarks had been expanded to include the fourth, charmed quark, it seemed that another such miracle, especially just in our proposed energy range, was too much to hope for.

What would we call the proposed machine? We needed a shorter, catchier label than the Cornell Storage Ring. Contemporary machines had been given acronyms like DORIS, SPEAR, PETRA and PEP. For a while it was open season on creative names. One of the wackiest I remember was suggested by Prof. Hywel White: **CORN**ell **CO**lliding **B**eams, or CORNCOB. Eventually, McDaniel ended the debate with CESR, the Cornell **E**lectron **S**torage **R**ing, pronounced like "Caesar". The name has weathered well, and has spawned others via Caesar's Egyptian connection: CLEO for the experimental detector and the collaboration that operates it, NILE for a computing project involving the CLEO data stream, and SUEZ for the data analysis main program.

4

The CLEO Proposal, 1975–1977

——————————————————— • ———————————————————

Besides a plan for the storage ring, we needed plans for two experimental detectors, one at each interaction point. The tradition at Cornell and other fixed target accelerator facilities had been to entertain proposals for experiments. The winners in the competition for approval would set up their apparatus, which would be torn down and dispersed as soon as the proposed measurements had been made. Over the years the experiments had become more complicated, the collaborations larger and the equipment more expensive. The idea had evolved at several labs that the most complicated apparatus would serve as a semipermanent, multipurpose, Laboratory-managed facility for the use of many experimenters in a long series of measurements. We decided that the south area was appropriate for such a facility, to be planned, built and exploited by a collaboration which would be open to all comers as long as the total number did not get too unwieldy. The smaller north area would be opened for competitive proposals by preformed collaborations.

So in 1975 a "South Area Experiment" study group started to form out of the Cornell 10 GeV Synchrotron user community, consisting of faculty, postdocs and graduate students from Cornell, Harvard, Rochester and Syracuse. Groups from Rutgers and Vanderbilt joined a little later, as well as individuals from Ithaca College (Ahren Sadoff) and LeMoyne College (David Bridges). We met under the chairmanship of Cornell's Prof. Al Silverman and started to consider the various options for detector technologies.

There was already a bewildering array of possibilities that had been considered by various summer study groups around the world, and some of the detector styles had even been built and used at Stanford, Hamburg, Frascati and Novosibirsk. First, there had been the nonmagnetic detectors with planar tracking chambers and shower hodoscopes. Then there were magnetic spectrometers along the lines of those used in some fixed target experiments. The latest was the Mark I solenoid-based detector with cylindrical tracking chambers. Other magnetic field configurations, longitudinal, transverse and toroidal, were being considered. There were serious limitations in every option; there was no detector that would excel in all respects. One had to make serious compromises between what was desirable for the physics capabilities and what one could expect to build with available resources.

It was assumed from the beginning that we would need a magnetic detector to achieve good momentum resolution for charged particles, and that the acceptance solid angle should be as near 4π steradians as practical. For a while I worked on the idea of a large open-gap magnet with the beam line parallel to the field direction and passing through a hole through the center of its poles. Although the produced particles over most of the solid angle range would be able to reach the detection devices without obstacles, the design suffered from the inconvenience of a very nonuniform magnetic field. Of the various configurations, the solenoidal style eventually won out. It promised a large acceptance solid angle for charged particle tracking without encumbrances, and the uniform magnetic field would simplify the track recognition and the momentum determination. However, the resolution would be poor for tracks at polar angles θ near zero and 180°. To get the best momentum resolution the solenoid coil would have to be big and expensive, and the other detector elements, particle identification, shower counters and muon detectors, would have to be even larger. To keep costs down, all but the tracking chambers would have to be outside the solenoid coil, and we would have to contend with the interactions of the particles as they passed through the coil. And because of their size, the outer detector elements would have to be made using low-tech, cheap technology. Still, it was probably the best choice.

The solenoid coil was a problem. We needed a high magnetic field for momentum resolution, but we had to minimize the thickness. The obvious solution seemed to be a thin superconducting coil. The TPC group at Berkeley was developing one for their detector at the new PEP storage ring at SLAC, so we assigned a Cornell postdoc, David Andrews, to the job of copying their design and getting us a 1.5 Tesla, 1 meter radius, 3 meter long solenoid. It was clear, however, that this was not going to happen quickly, so we decided to build a temporary conventional 0.5 Tesla coil as well.

For the tracking chamber, we chose a cylindrical drift chamber of 17 layers of square cells, alternating axial and 3° slanted wire layers to get stereo information. This was inspired by the drift chamber for the Mark II detector, which was being planned at that time to replace the Mark I at SPEAR; Prof. Don Hartill had just spent a sabbatical year at SLAC participating in the design. In the space between the drift chamber and the beam pipe we planned a proportional chamber with cathode strips to measure the track z coordinate parallel to the beam line. Each component of the detector got a two-letter mnemonic to identify it in the software, and was usually referred to by these two letters in the local jargon. The big cylindrical drift chamber was "DR" and the inner z chamber was "IZ".

Outside the coil and inside the iron of the flux return interleaved with planes of drift chambers for "MU" muon detection, the detector was to be arranged in octants, each octant consisting of a separate trapezoidal box containing, in outward order, the "OZ" planar drift chamber to track each charged particle after it had passed through the coil, a device for particle identification, a plane of scintillation counters "TF" for triggering and time of flight measurement, and the "RS" shower detector array of alternating lead sheets and proportional tubes. To extend the solid angle for photon detection we also covered the ends of the solenoid with similar "ES" shower detectors mounted on the iron poles, and another set of such detectors "CS" mounted at the ends of the octant modules.

While these components were agreed on with a minimum of controversy, we were not able to agree on a scheme for high momentum charged particle identification. Measuring pion, kaon, proton, electron and muon masses was possible by combining the measurement of momentum p by curvature in the magnetic field with the measurement of velocity β by time of flight ($m = p\sqrt{1 - \beta^2}/\beta$), but only up to $\beta \approx 0.95$ above which their flight times became indistinguishable. Thus pions and kaons of the same momentum could not be separated above 800 MeV/c using just momentum and time of flight. Two other physical processes that depend on particle velocities are energy loss by ionization dE/dx and the Cherenkov effect, and there were partisans for each. Frank Pipkin and the Harvard group proposed to use high pressure gas Cerenkov counters. They chose the gas pressure to give an index of refraction n that would yield a threshold value, $\beta_{\min} = 1/n$, such that for momenta in an interesting range pions would count and kaons would not. Provided we could solve the light collection problems, achieving this limited goal would be straightforward. The more ambitious and risky dE/dx measuring scheme was championed by Professors Panvini, Csorna, and Stone of Vanderbilt and Prof. Richard Talman of Cornell. They promised to distinguish pions and kaons cleanly to momenta slightly higher than possible with time of flight, and also provide some minimal separation in the dE/dx relativistic rise region, $p > 1.2$ GeV/c. Moreover, since one would have a β measurement rather than a signal only when $\beta > \beta_{\min}$, the device would help in separating other particle species, for example, p versus K and e versus π. The actual performance, however, would depend critically on the resolution in the measurement of dE/dx; could we achieve 6% r.m.s.? Without a dictator in charge of CLEO, we argued and dithered and were not able to come to a decision. Eventually we agreed to equip two octants with high pressure "CV" gas Cerenkov counters and two octants with "DX" proportional wire chambers, so we could try them both out for a while before committing all the octants. Later, Steve Olsen of Rochester suggested building simple low pressure gas Cerenkov detectors to fill the empty octants temporarily and help distinguish electrons and pions by velocity threshold. While the inner part of the detector demonstrated general agreement within the collaboration, the outer components, with three different kinds of particle identification, reflected the conflicts.

The collaboration needed a snappy name. Other collaborations tended to have acronyms like DASP, LENA, TASSO, JADE, DELCO, or were named for personalities like PLUTO, or had unpronounceable initials like TPC or HRS. It was a graduate student, Chris Day, who suggested "CLEO", short for Cleopatra. To make it into an acronym no one could think of a better set of words than "Cornell's Largest Experimental Object" or "...Operation" or "...Organization", so we decided not to make it an acronym — it is just a name. By now, people have stopped asking what it stands for.

Meanwhile, plans for the north area experiment were being resolved. A call for proposals was sent out, and on 15 February 1978 the Program Advisory Committee met to consider the submissions from three groups: Columbia + Stony Brook, Chicago + Princeton and the University of Massachusetts. All three involved compact nonmagnetic detectors emphasizing calorimetry, quite complementary to the CLEO detector design. The winning entry was the sodium iodide and lead–glass array proposed by the group from Columbia and Stony Brook with Leon Lederman as spokesman. Lederman left almost immediately afterwards to become director of Fermilab and the direction of the collaboration, called CUSB, was taken over by Paolo Franzini and Juliet Lee-Franzini.

5

Approval and Construction, 1977–1979

In the summer of 1975, a subpanel of the US High Energy Physics Advisory Panel (HEPAP) met at Woods Hole on Cape Cod, under the chairmanship of Francis Low, to advise ERDA, the predecessor of the DOE, on new high energy physics facilities. Even though the NSF was not bound to follow the recommendations of HEPAP or its subpanels, it was important for us to get an endorsement for CESR. There were four project proposals on the table: the PEP 18 GeV e^+e^- ring for SLAC, the ISABELLE 200 GeV pp collider for Brookhaven, the "Energy Doubler/Saver" superconducting-magnet proton synchrotron ring for Fermilab, and the CESR proposal from Cornell. The previous year's subpanel under the chairmanship of Victor Weisskopf had already endorsed PEP for construction and had recommended continued R&D efforts for ISABELLE and the Fermilab ring, however the PEP project had not yet received funding.

I was a member of the Low Subpanel, but I was in the minority on most issues. Most of the members felt that getting PEP launched was the first priority, and that it was therefore impolitic to recommend a second cheaper e^+e^- collider as well. So in the final priority list CESR came in fourth. PEP was reaffirmed for construction, ISABELLE and the Energy Doubler/Saver were recommended for increased R&D funding, and although there was some faint praise for the Cornell proposal, the report said, "we do not consider the construction of a second electron–positron colliding beam facility as one of our highest priorities."

This was a dark moment for CESR, but McDaniel was undaunted. Once the PEP project obtained government approval a few months later, he contacted individually the members of HEPAP and the Low Subpanel and got many of them to support the CESR proposal. Assurances were given that the experimental program would have greater participation by non-Cornell groups. Most importantly, we were

Fig. 10.   Joe Kirchgessner and the body of the Mark I CESR 14-cell rf cavity.

extremely fortunate in getting the enthusiastic support of Al Abashian and Marcel Bardon at the NSF Physics Division [11]. But as Abashian recalls [7], "History notes that the birth of Julius Caesar was a particularly difficult one for his mother and ultimately necessitated the adoption of radical measures." To keep a low profile and avoid alarming the watchdogs at the OMB (Office of Management and Budget) with the prospect of a new facility, the project was officially relabeled a *conversion* of the existing synchrotron facility. During 1976 and 1977 the NSF Physics Division provided enough funds for a vigorous program of prototyping and firming up the design. We were even able to lengthen and upgrade the linac for positron production, using sections from the decommissioned Cambridge Electron Accelerator. Finally, late in 1977 the official approval came from the National Science Board (the governing board of the NSF), and $20.6 million was eventually awarded for construction of CESR and the CLEO detector.

It is ironic that of the four facilities rated by the Low Subpanel, CESR was the first to come into operation, and by the 1990s the two with the lowest priority, CESR and the Fermilab ring, now called the Tevatron, were the only two producing physics results.

McDaniel put Maury Tigner in charge of CESR construction, while Al Silverman led the CLEO effort. The schedule was ambitious and called for first beam trials on 1 April 1979, less than two years away. The various systems — magnets, rf (Fig. 10), injection, vacuum, controls — all had challenging performance goals. Many of the Cornell experimenters joined the CESR effort. For example, I worked on the synchrotron to storage ring injection beam transport while my involvement with the

Fig. 11. Elsa Adrian and Gino Melice stringing the wires of the CLEO DR1 drift chamber.

CLEO detector was in building the luminosity monitor and writing the trackfinding program for the DR. For the final pulsed bending magnet in the injection beam line I needed a source of thin iron sheeting for the laminated magnet core. I was delighted to find in storage the old flux-bar iron from the 300 MeV synchrotron. It was ideal. So CESR now contains a recycled piece of the original Cornell machine.

The CESR construction program was extremely well organized under Tigner, and plowed along relentlessly in spite of difficulties. CLEO, on the other hand, was a free association of individual, far-flung university groups accustomed to working independently on much smaller projects with little or no time pressure. It needed all of Silverman's skills at negotiating and coaxing to keep it on any semblance of schedule. The various components of the detector were parceled out to the university groups, as in the table below. Typically, each group took responsibility also for the readout electronics and software associated with its detector component. Figures 11 and 12 show the stringing of the DR chamber and the assembly of the detector in the interaction region pit.

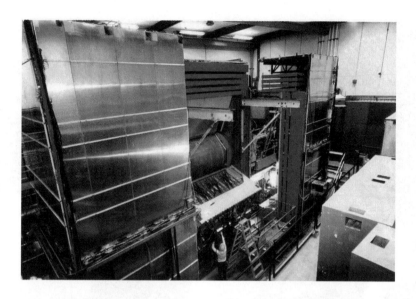

Fig. 12. The CLEO-1 detector partially installed in the L-0 pit. The sheet metal surfaces are the MU drift chambers. One can see the solenoid coil and one of the octant modules containing OZ, TF, DX, RS and OZ.

LM	small angle luminosity detector	Cornell
IZ	inner z proportional chamber	Syracuse
DR	main cylindrical drift chamber	Cornell
	solenoid coil	Cornell
OZ	outer planar drift chambers	Syracuse
CV	high pressure gas Cerenkov	Harvard
	low pressure gas Cerenkov	Rochester
DX	dE/dx gas proportional chambers	Vanderbilt
TF	time of flight scintillators	Harvard
RS	octant shower detector	Rutgers
ES	pole-tip shower detector	Harvard
CS	octant-end shower detector	Harvard
	magnet yoke	Rochester
	additional iron for muon filter	Harvard
MU	muon drift chambers	Rochester

By the end of March 1979 the installation of the essential CESR components in the tunnel was just about complete (see Fig. 13, top). On the evening of April 1, Nari Mistry and a few helpers were pushing to get the vacuum system in shape for the scheduled beam turn-on. Nari was a Cornell Research Associate who had got his degree from Columbia University working with Leon Lederman on the experiment that established that the muon neutrino and the electron neutrino were distinct particle species. Over the years, Nari has made many contributions to CESR and

Fig. 13. (top) View of the CESR tunnel showing the injector synchrotron on the left, the storage ring on the right, and Boyce McDaniel in the middle. (bottom) Celebration in the CESR control room on the occasion of the first storage of a positron beam. Seated is Maury Tigner. Behind him are Joe Kirchgessner, Gerry Rouse, Chuck Chaffey, Raphael Littauer, Ernie vonBorstel, Boyce McDaniel, Bob Siemann, Ron Sundelin, Mario Gianella, Nari Mistry, Dave Rice, Al Silverman, Dave Andrews, Gordon Brown, an unknown, Dave Thomas, Dave Morse, C.O. Brown, Ken Tryon, Jim Fuller, Karl Berkelman and Peter Stein.

CLEO, but perhaps his most significant accomplishment is that he became the world's leading guru on synchrotron vacuum systems. As an uncompromising perfectionist, he was well suited to the vacuum effort.

As I recall, the first electrons were injected into CESR the next morning, on what would have been April 2 if we had not stopped the clock. The first stored electron beam was achieved on Friday the 13th. The next several months were spent establishing electron beams and then positron beams in the storage ring [8] (Fig. 13, bottom). Beam trials were held in the evenings so that installation work could go on during the days, especially for the CLEO detector. It was on August 14 that we had the first measurable colliding beam luminosity. By October the luminosity was enough to schedule the first experimental run of the CLEO detector. It was still missing two octants of outer detector and half of the muon chambers, but it was enough to start looking for electron–positron elastic scattering and annihilations into hadrons.

Looking back now, it seems to me that the building of CESR in less than two years well within the modest $20 million budgeted was a major achievement. There were many who had said we could not do it, and certainly not for that price. In fact, we had beaten the PEP turn-on time by about a year, and they had started earlier.

6

First Data, 1979–1980

———————————— • ————————————

In the original 1975 CESR proposal we had said, after a discussion of the charm quark threshold, that "present theory is quite inadequate to predict whether further hadronic degrees of freedom exist at even higher energies, ... we do not know at what energies such new thresholds occur." But in a later version of the proposal, written in October 1976, there is a prophetic section written by Cornell theorist Kurt Gottfried in which he speculates on the consequences of a threshold for a hypothetical heavier quark in the CESR energy range [5]. He includes a figure labeled "The spectrum of $Q\bar{Q}$ bound states for a heavy quark having a mass of 5 GeV".

It turned out to be an amazingly accurate prediction of the upsilon bound state energy levels and the $\pi\pi$ and γ transitions among them, a year before the discovery in 1977 by Lederman's group [9],[12] at Fermilab of a peak in μ-pair production at 9.4–10.4 GeV mass. In a replay of Ting's finding the J (or ψ) at the charm quark threshold, the upsilons were immediately interpreted as the quark–antiquark bound states of a heavier, fifth quark called the bottom quark or b. This is a good place to take time out to detail what I call "the b quark serendipity", although many of the following facts would be revealed only years later in the life of CESR and CLEO.

- The new, fifth quark had a mass that matched the CESR energy range for good beam luminosity. The masses of the $b\bar{b}$ bound states and the range immediately above threshold for producing pairs of mesons containing b quarks fell between 9.4 and 10.6 GeV, energies that were too high for SPEAR and too low for PEP, for example, but were nearly optimal for CESR.
- At only 22 MeV above the $B\bar{B}$ meson production threshold, there was a resonance, the $\Upsilon(4S)$ state. Its decay was therefore phase space suppressed,

making the resonance unexpectedly narrow, causing it to stand out clearly above the background of e^+e^- annihilation into lighter quark states. Moreover, the 22 MeV was not enough energy to produce any other mesons along with the B's, so the final states of the $\Upsilon(4S)$ decay were unexpectedly clean.

- The b quark is lighter than its weak doublet partner, the t ("top") quark. Energy conservation then forces the b to decay outside its doublet via the off-diagonal elements of the Cabbibo–Kobayashi–Maskawa (CKM) matrix V_{ij} that specify the weak couplings between charge 2/3 and charge $-1/3$ quark species.

- The most likely decay channel, to the c quark, is parametrized by V_{cb}, which turned out to be quite small, $|V_{cb}| = 0.04$, much smaller than the analogous Cabibbo matrix element, $|V_{us}| = 0.23$. This leads to an unexpectedly long 1.6 ps lifetime for the B meson, long enough for a B with modest kinetic energy to have a measurable decay distance, and long enough for the B^0 to oscillate to its $\overline{B^0}$ antiparticle.

- The alternate decay route, to the u quark, has $|V_{ub}| = 0.004$ — a very small but *nonzero* matrix element. Because V_{ub} is so small, the rates for charmless decays of B mesons are sensitive to higher order weak processes and possibly to new physics beyond the Standard Model. It is the fact that V_{ub} is nonzero that allows for the Kobayashi–Maskawa mechanism of CP violation in weak decays.

- The 174 GeV mass of the t quark gives it plenty of phase space to decay to its partner b plus a real W boson. This, combined with the expectation that $V_{tb} \approx 1$, implies a lifetime of the order of 5×10^{-25} second, which is much too short to allow the t to form any bound states. That leaves the $b\bar{b}$ bound states as the most ideally nonrelativistic strongly bound quarkonium system. Upsilon spectroscopy is our best laboratory for testing our understanding of strong binding forces.

These were all very lucky breaks for CESR and CLEO, guaranteeing decades of productive heavy quark physics research. And now, after this peek ahead, we return to the early days of CESR and CLEO.

The Fermilab experiment [9] could not resolve the three bound states, although the shape of the mass peak clearly favored more than one. Lederman and collaborators even claimed the presence in their data of a third upsilon state not obvious to the naked eye, but I doubt if they would have had the courage to do so if the "Cornell" potential model of Gottfried *et al.* had not already predicted a third level. In a replay of the Adone post-discovery of the ψ, the DORIS e^+e^- storage ring at DESY was quickly beefed up to run at an energy high enough so that in early 1978 the PLUTO and DASP detectors could locate the lowest upsilon state and later in the year the DASP and LENA detectors could see the first two states, then called the Υ and Υ'. DORIS would not be able to reach the third state until several years later, after they had converted from two ring operation to a single ring machine running in the SPEAR single bunch mode.

So at the time that CESR came into operation the questions were:

- Was there really a third upsilon bound state resonance?
- Would the $b\bar{b}$ spectroscopy look like $c\bar{c}$ spectroscopy, that is, was the strength of the binding potential independent of quark flavor?
- Was there a threshold for "open-b", that is, for the production of B meson pairs, and if so, what energy did it occur at?
- Would there be a quasibound resonance just above threshold, analogous to the $\psi''(3770)$, where the cross-section for $B\bar{B}$ would be enhanced?
- Could one find evidence for a new quark flavor by seeing leptonic decays above the open-b threshold?
- Would the value of $R = \sigma(e^+e^- \to \text{hadrons})/\sigma(e^+e^- \to \mu^+\mu^-)$ well above threshold confirm that the b had charge $-1/3$?
- Would there be a t quark, partner of the b, and if so what was its mass?

In November we started measuring the total hadronic cross-section as a function of beam energy in small steps around where we expected to find the upsilon resonances. It took a few days to get the trigger and data readout working reliably. Also, since no one had any confidence in my tracking and hadronic event selection software (least of all I), we tried several independent schemes for determining the cross-section, for example, scanning the pictorial display of the tracks by eye (Fig. 17) and counting the ones that appeared to be beam–beam annihilations, or picking the events on the basis of shower energy. To the accuracy we needed, the various analysis methods agreed, and there was the CUSB experiment for confirmation, too. At the luminosity CESR had achieved by then, about $10^{30}/\text{cm}^2\text{sec}$, the hadronic event rate off resonance was about 10 per hour. Once we started the energy scan, it did not take long to find the first resonance. When we attained the right energy, the rate rose to about 1 per minute. This calibrated the CESR energy scale relative to that of DORIS. Since DORIS had already measured the energy difference between the Υ and the Υ', CLEO and CUSB were able to locate the second resonance almost immediately.

McDaniel had the idea of announcing CESR to the world by a Laboratory holiday greeting card, showing the data for the two resonances, cross-section versus energy. By the time he was ready to get it printed up, we had found the third resonance, thus confirming its existence and making the first accurate measurement of its mass. So the Υ'' data were added to the card (see Fig. 14).

For the next few months while continuing to take data, CLEO and CUSB worked out the efficiencies and corrections, wrote their papers announcing the "Observation of Three Upsilon States", and submitted them back to back to *Phys. Rev. Lett.* on 15 February 1980 (see Appendix, Table V). The first CLEO paper (see Fig. 15) had

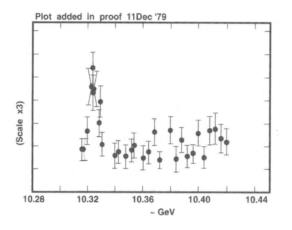

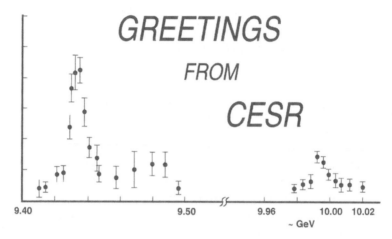

Fig. 14.   1979 holiday greeting card from CESR.

73 authors from 8 institutions. Twenty-two of the authors and 6 of the institutions were still in CLEO fifteen years later.

By now CLEO was an established collaboration with elected officers, regular monthly meetings and written minutes. Of all the large collaborations in high energy physics, it is probably the most democratic. Collaboration policies, officers, physics goals, what to publish, whom to admit for membership, are all decided by majority vote of all the members — faculty, post-docs, and students — at the regular meetings. Elections are held every spring and the new officers take up their positions in the middle of the summer (see Appendix, Table 3). Opportunities to represent CLEO at conferences are pooled and assigned to CLEO members by a broadly representative committee so that most members get a chance to speak for the collaboration. Although the CLEO governance structure can sometimes be rather cumbersome and the collaboration often finds it difficult to make decisions,

VOLUME 44, NUMBER 17   PHYSICAL REVIEW LETTERS   28 APRIL 1980

Observation of Three Upsilon States

D. Andrews, K. Berkelman, M. Billing, R. Cabenda, D. G. Cassel, J. W. DeWire, R. Ehrlich,
T. Ferguson, T. Gentile, B. G. Gibbard,[a] M. G. D. Gilchriese, B. Gittelman, D. L. Hartill,
D. Herrup, M. Herzlinger, D. L. Kreinick, D. Larson,[b] N. B. Mistry, E. Nordberg,
S. Peggs, R. Perchonok, R. K. Plunkett, J. Seeman, K. A. Shinsky, R. H. Siemann,
A. Silverman, P. C. Stein, S. Stone, R. Talman, H. G. Thonemann, and D. Weber
Cornell University, Ithaca, New York 14853

and

C. Bebek, J. Haggerty, J. M. Izen, R. Kline, W. A. Loomis, F. M. Pipkin,
W. Tanenbaum, and Richard Wilson
Harvard University, Cambridge, Massachusetts 02138

and

A. J. Sadoff
Ithaca College, Ithaca, New York 14850

and

D. L. Bridges
Le Moyne College and Syracuse University, Syracuse, New York 13210

and

K. Chadwick, P. Ganci, H. Kagan, F. Lobkowicz, W. Metcalf,[c] S. L. Olsen, R. Poling,
C. Rosenfeld, G. Rucinski, E. H. Thorndike, and G. Warren
University of Rochester, Rochester, New York 14627

and

D. Bechis, G. K. Chang,[d] R. Imlay,[c] J. J. Mueller, D. Potter, F. Sannes, P. Skubic, and R. Stone
Rutgers University, New Brunswick, New Jersey 08854

and

A. Brody, A. Chen, M. Goldberg, N. Horowitz, J. Kandaswamy, H. Kooy, P. Lariccia,[e]
G. C. Moneti, and R. Whitman[f]
Syracuse University, Syracuse, New York 13210

and

M. S. Alam, S. E. Csorna, R. S. Panvini, and J. S. Poucher
Vanderbilt University, Nashville, Tennessee 37235
(Received 15 February 1980)

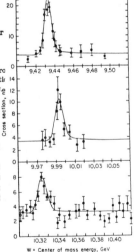

Three narrow resonances have been observed in e^+e^- annihilation into hadrons at energies between 9.4 and 10.4 GeV. Measurements of mass spacings and ratios of the pair widths support the interpretation of these "Υ" states as the lowest triplet S-states of the $b\bar{b}$ quark-antiquark system.

PACS numbers: 13.65.+i

We report here on the first results from the CLEO detector at the Cornell Electron Storage Ring (CESR). CLEO is a magnetic detector built around a 1.05-m-radius, 3-m-long solenoid coil producing a magnetic field parallel to the beams (see Fig. 1). Charged particles are observed and their momenta measured over a solid angle of 0.90×4π sr in a cylindrical ... ing most of the field volume ... cal proportional chamber ... ing the beam pipe provides ... along the beam axis. Outs ... length thick aluminum sol ... tion counters 2.2 m from t

Fig. 15. The first CLEO publication.

I believe that the tradition of equality has done wonders for the group spirit and loyalty of the membership, especially the younger members. All share in the decisions and no one feels exploited. Originally some collaborators feared that Cornell might be too dominant in CLEO; for example, Frank Pipkin of Harvard had written to McDaniel, "It strikes me that the most sensible user arrangement is one in which the groups are composed in part from Cornell people and in part from outside users with sufficient balance in talent and contribution of apparatus that each needs the others in a very real way. One should avoid the SLAC model in which the inside component always has the upper hand." I believe that over the years we have managed to allay those fears; in fact, as the collaboration has acquired new members, the relative weight of the outside groups has steadily increased. The collaboration owes a lot to Al Silverman for guiding it in the formative years and establishing the effective and collegial CLEO traditions.

Although there was a lively CLEO interest in the upsilon bound states, attention moved to the search for B mesons, the bound states of b quark and $\bar{u}$ or $\bar{d}$ antiquark. The nonrelativistic models of the $b\bar{b}$ binding potential could predict the relative masses of what we now called the $\Upsilon(1S)$, $\Upsilon(2S)$ and $\Upsilon(3S)$ bound states, but they were not able to predict masses for $b\bar{u}$ and $b\bar{d}$ mesons, that is, locate the threshold energy at which the $b\bar{b}$ produced in the e^+e^- annihilation would appear as B^+B^- or $B^0\overline{B^0}$. What would be the cross-section above threshold, and would we be able to see it? There was the possibility that nature might be kind and give us an almost bound $\Upsilon(4S)$ resonance just barely above threshold so that (a) its decay would be inhibited enough for it to be narrow (and therefore tall) peak in an energy scan of the e^+e^- annihilation rate, and (b) so that the resonance would decay only to $B +$ anti-B pairs without even an additional pion. But since the spacing of the upsilon levels was several hundred MeV ($M_{3S} - M_{2S} = 332$ MeV for example), it seemed too much to hope for.

We started scanning the annihilation cross-section between 10.3 and 10.6 GeV e^+e^- total energy. By April we knew we had hit the jackpot. Nature had been incredibly kind and had given us a beautifully high and narrow $\Upsilon(4S)$ resonance just 22 MeV above $B\bar{B}$ threshold. At the resonance energy one in every four hadronic events was $e^+e^- \rightarrow B^+B^-$ or $e^+e^- \rightarrow B^0\overline{B^0}$. This energy, 5.29 GeV per beam, would be the energy at which CESR would run for most of its life.

If the Standard Model was now assumed to be based on 6 quarks and 6 leptons with the quarks coming in three colors, the b quark should decay to a c or u quark plus a W^-, and naive counting rules would imply that W^- should materialize 1/9 of the time to $e^-\bar{\nu}_e$ and 1/9 of the time to $\mu^-\bar{\nu}_\mu$. Thus one would expect to see e's and μ's produced at the $\Upsilon(4S)$ at rates much higher than in the continuum on either side of the resonance. Ed Thorndike and the Rochester group confirmed that the lepton rates were compatible with the counting rules. This established the existence of the new quark flavor, and can be considered the discovery of the B meson. By the end of 1980 CLEO had submitted four important papers (see Appendix, Tables 4, 5, 11) that set the course of CESR physics for the next decades.

- Observation of three upsilon bound states.
- Observation of a fourth, wider upsilon state in e^+e^- annihilations.
- Evidence for new-flavor production at the $\Upsilon(4S)$.
- Decay of b-flavored hadrons to single-muon and dimuon final states.

Al Silverman's review talk at the 1981 Lepton Photon conference in Bonn is a nice summary of the early CLEO and CUSB results.

7

The CESR-II Blind Alley, 1980–1983

Flushed with their success in building CESR and turning it into a productive upsilon physics facility, Maury Tigner and the Cornell accelerator physics crew were looking for new worlds to conquer. The superconducting rf cavity development effort that Maury had started back in the mid 1970s as a way of increasing the synchrotron energy was showing promise, but not for CESR. At the CESR beam energy and intensity the rf power dissipated in the cavity walls was not high enough (compared with the power radiated by the beam) to make the superconducting alternative economically attractive. But if high enough field gradients could be obtained, superconducting cavities might significantly reduce the size and cost of a very high energy e^+e^- ring.

In 1980, the CERN UA1 and UA2 experiments were being turned on at the SPS collider, with the major goal of finding the $W^\pm$ and Z^0 weak intermediate vector bosons in $\bar{p}p$ collisions. CERN and SLAC were proposing to build the LEP and SLC e^+e^- colliders to exploit the physics at the Z^0 resonance at about 91 GeV in the center of mass. LEP was to be a 26.7 km ring costing over half a billion dollars, and the SLC would depend on an untried linear collider scheme with rather risky prospects for luminosity. It seemed to Maury and others that here was an opportunity. They drew up in May 1980 a "Design Study Proposal" for a 50-on-50 GeV e^+e^- ring with a circumference of 5.485 km, luminosity $3 \times 10^{31}/\text{cm}^2\text{sec}$ (the same as for LEP), to cost about $150 million. It was actually just an R&D proposal, since the project needed additional design work and superconducting cavity development. At first, a site not far to the east of the existing CESR ring was considered; later another site was projected northeast of the Tompkins County Airport.

The Lab hosted workshops in November 1980, and in January and April 1981, to advance the machine design, discuss the experimental detectors, and generate

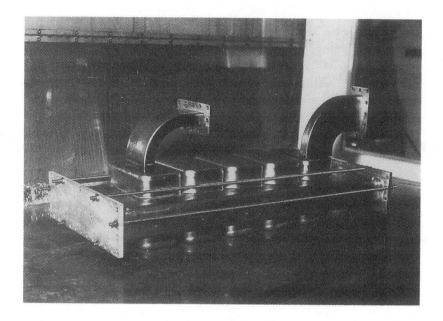

Fig. 16.   A five-cell muffin-tin niobium cavity.

interest in the high energy physics community outside Cornell. Several thick reports on "CESR-II" were produced.

The superconducting rf cavity work had advanced to the stage of building a prototype accelerating system suitable for a test in CESR, which took place in 1982. There were problems, however. Maury had invented a clever "muffin tin" structure for the cavities (see Fig. 16), that was relatively cheap to build and prevented synchrotron radiation from striking the cavity walls. But the resonant multiplicative emission and reemission of electrons from the cavity walls (called multipacting) limited the achievable field gradients in spite of measures taken to suppress the effect. In Europe they were having much better success with axially symmetric ellipsoidal cavities, which channeled the emitted electrons to lower-field regions where they would not multiply when they struck the cavity walls.

The NSF encouraged the superconducting rf development effort, but was cool to the idea of spending \$150 million on a Z^0 factory. HEPAP formed another subpanel in August 1981, chaired by George Trilling, to consider the various accelerator proposals: the Brookhaven ISABELLE pp collider, the Fermilab Tevatron, and the e^+e^- SLAC Linear Collider (SLC). Since it was not yet a real construction proposal, CESR-II got only passing mention in the subpanel report. The subpanel did recommend, however (I was a member), that the SLC project be funded, the major motivation being the potential advance in accelerator technology. Most of us were not completely convinced that the SLC would succeed in making enough luminosity to compete favorably with LEP — and we were right.

Now that LEP and SLC were practically launched, the prospects for NSF funding for CESR-II got even dimmer. The enthusiasm was lost. Apparently stymied in his quest for a great leap forward at Cornell, Maury Tigner looked around for another challenge. In April 1983 he convened at Cornell the workshop that launched the design work for the Superconducting Super Collider, and in 1984 he left Cornell to head the SSC Central Design Group at Berkeley. At various times, others like Professors Murdoch Gilchriese and Richard Talman joined the effort. McDaniel took on the chairmanship of the SSC Board of Overseers and retired as Director of LNS in 1985. I succeeded him as the Laboratory's fourth Director. Recall that the first three were Robert Bacher in 1946, Robert Wilson in 1947 and Boyce McDaniel in 1967.

One consequence of the death of the CESR-II Z-factory idea was the loss of mission for the superconducting rf (SRF) group in the Lab. Over the years the group had grown under Tigner's direction to include Ron Sundelin, Joe Kirchgessner, Hasan Padamsee, Peter Kneisl, Charles Reece and Larry Phillips — all Research Associates or Senior Research Associates. They had turned the old Newman Lab synchrotron area into an impressive microwave and superconductivity research and development complex. Looking around for a mission for SRF, Sundelin convinced the designers of the CEBAF (Continuous Electron Beam Accelerator Facility) 4 GeV electron accelerator project for nuclear physics research in Newport News, VA that what they needed was a CW, superconducting, recirculating, linear accelerator. While CEBAF was getting approved and set up as a laboratory, Sundelin got DOE funding to turn the latest model SRF ellipsoidal cavities, recently tested in CESR, into industrial prototypes for the CEBAF accelerator. When this work was completed in 1987, most of the SRF group, all except Padamsee and Kirchgessner, left to work for CEBAF. I had a hunch that SRF would eventually prove useful for the future Lab program, and since the NSF wanted to preserve the Cornell effort in SRF, I decided against McDaniel's advice to replace, at least partially, the personnel losses and keep the research effort alive. For the next several years the SRF group under Padamsee's direction concentrated on basic studies of the phenomena that limited attainable field and Q, and on development of cavity structures that would be an economical possibility for a future TeV energy linear electron–positron collider. The group also studied the performance of Nb_3Sn and high-T_C superconductors in microwave fields. In later years, my hunch proved correct and SRF cavities became an important component of a high luminosity CESR upgrade.

Another idea for a future direction for the Laboratory surfaced in 1983. The Major Materials Facilities Committee, named to advise the President's Science Advisor, recommended the construction of a dedicated 6 GeV Synchrotron Radiation Facility. To many people this seemed like a natural for Cornell. In April 1984, McDaniel and CHESS Director Boris Batterman circulated a prospectus for such a $66 million ring in a separate tunnel near CESR. Bob Siemann also got enthusiastic about it. It would provide a guaranteed future for the accelerator physicists and other lab employees, even though it would have no interest for the high energy physicists. After a meeting of all concerned, however, it was clear that the majority of the accelerator physicists also had no interest in working on a machine that did not serve high energy physics. Later, Argonne National Laboratory built the 6 GeV Advanced Photon Source.

Meanwhile, by the mid 1980s "b quark serendipity" was beginning to have its effect. b quark physics at CESR had turned out to be more exciting than anyone had anticipated, and the future of the Laboratory seemed to be pointed in that direction. The atmosphere of doom was beginning to dissipate.

8

The CLEO-1 Years, 1981–1988

———————————————— • ————————————————

The users of CESR in the 1980s were CLEO at the south interaction point, CUSB in the north, CHESS using the X-ray beam lines to the west of CLEO, and the Cornell accelerator physics faculty, staff and graduate students.

In the early days of CESR, a separate organization was set up to manage the exploitation of synchrotron radiation for Cornell and outside users interested in the X-ray capabilities. The Cornell High Energy Synchrotron Source (CHESS) was under the direction of Prof. Boris Batterman of the Department of Applied and Engineering Physics, with Prof. Neil Ashcroft of the Physics Department as Associate Director. Three beam lines (called A, B and C) were set up to exploit the intense radiation from the electrons passing through the CESR hard bend magnets just to the west of the south intersection region. Except for occasional brief periods, no longer than a month, the synchrotron radiation program was to be parasitic to the high energy physics running. In spite of this and competition from Brookhaven and other dedicated facilities with many more beam lines, CHESS always had a very active program with many loyal users from Cornell, outside universities, government labs and industry. Until the arrival of the Grenoble and Argonne rings in the mid 1990s the CHESS beams had the highest available energies, and as the luminosity of CESR was increased for high energy physics, CHESS continued to lead the world in X-ray intensities.

Although there was always the potential for friction because of the conflicting requirements for beam conditions and other resources, the symbiosis between CHESS and LNS worked remarkably smoothly. Each program obviously benefited from the presence of the other. CHESS got their beams for free, while the goodwill that LNS gained at the NSF from a broad user community was often crucial in getting support for CESR funding.

The Columbia–Stony Brook (CUSB) detector in the north area had photon energy resolution clearly superior to that of the CLEO-1 detector, and was ideally suited for the study of radiative transitions among the upsilon bound states. Both detectors could measure total annihilation cross-sections well, and therefore competed in bump-hunting for $^3S_1$ vector $Q\bar{Q}$ resonance states. Because of its magnetic field, the CLEO detector was superior for the study of the $\pi^+\pi^-$ transitions, like $\Upsilon(2S) \to \pi^+\pi^-\Upsilon(1S)$, but the 4-prong events in which the final Υ decayed to a lepton pair were so obvious that CUSB could also identify them cleanly without momentum measurement. CUSB was a small, close-knit, dedicated group that excelled in making the most of a limited detector, a very confined experimental area and scarce resources.

Most of the CLEO experimenters were more interested in the weak interactions of B mesons than in the strong interaction physics of the bound state resonances. So whenever the time came each year or so to present their requests to the Program Advisory Committee, CLEO asked for running time on the $\Upsilon(4S)$ resonance above $B\bar{B}$ threshold, and CUSB asked for time on the narrower $\Upsilon(2S)$ or $\Upsilon(3S)$ resonance. The Committee consisted of six to eight high energy physicists from other US laboratories and universities appointed by the Lab director in consultation with the NSF Program Officer. After hearing reports from CESR accelerator physicists and from CLEO and CUSB, as well as the proposals for future running, the PAC would award CLEO typically two-thirds of the priority and CUSB the remainder, but of course both experiments ran all the time.

From 1981 through to 1986, when the drift chamber was replaced, the CUSB priority runs resulted in 13 published CLEO papers on the Υ bound states: one on the total cross-sections, four on the $\Upsilon(2S$ or $3S)\to \pi^+\pi^-\Upsilon(1S)$ transitions (see Fig. 17), three on the dilepton decays of the upsilons, one on the $\Upsilon(2S)$ radiative decay to the $\chi_b(1P)$ states detecting pair-converted photons in the drift chamber, and four on radiative $\Upsilon(1S)$ decays (see Appendix, Table 5). Many of these were paralleled by CUSB papers based on the same CESR running. Most of the results confirmed the expectations from the potential models; that is, $b\bar{b}$ spectroscopy repeated the $c\bar{c}$ spectroscopy that had been worked out mainly at SPEAR a few years before. The pattern of energy levels and the kinds of transitions between them were similar, implying that the carrier of the strong force was flavor blind and that of the effective quark–antiquark binding potential $V(r)$ in the relevant range of separation distance was intermediate between Coulomb-like $(1/r)$ and linear in r. One new feature was the presence of three $^3S_1$ $b\bar{b}$ bound states below $B\bar{B}$ threshold, instead of the two $c\bar{c}$ states below the analogous $D\bar{D}$ threshold. Thus the Υ system has about double the number of states and transitions. With more energies and rates to measure, one could test the theory better and constrain its parameters more tightly. Also, with a much more massive quark the nonrelativistic theory had a better chance of being numerically reliable.

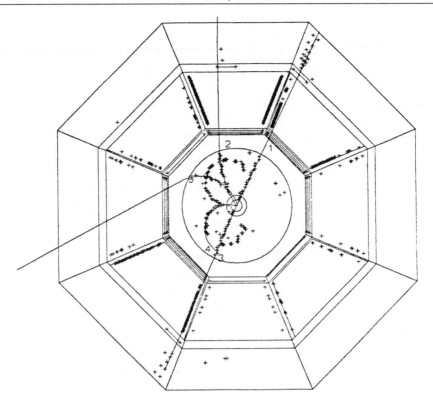

Fig. 17. Computer display of a candidate for $\Upsilon(2S) \rightarrow \gamma\chi_b(1P)$, $\chi_b(1P) \rightarrow \gamma\Upsilon(1S)$, $\Upsilon(1S) \rightarrow e^+e^-$ in the CLEO-1 detector. Each photon converts to an e^+e^- pair.

The one unanticipated result was the double-hump shape of the dipion effective mass spectrum in $\Upsilon(3S) \rightarrow \pi^+\pi^-\Upsilon(1S)$. It prompted speculations about $\pi\pi$ interactions and $\Upsilon\pi$ interactions, and still lacks a universally accepted explanation. Presumably, because it is hard to calculate long range phenomenon and is not a sign of something basically wrong with QCD.

Our competition in this era came from the DORIS e^+e^- storage ring at DESY, upgraded in energy after 1982 to reach as high as the $\Upsilon(4S)$ and equipped with the ARGUS and Crystal Ball detectors. ARGUS was a solenoid-based magnetic detector, at least as good as CLEO-1 and perhaps better, built and operated by the German, Russian, US, Canadian group that had taken over the DASP detector from its builders in 1978. The Crystal Ball, a mainly spherical array of sodium iodide scintillators, was optimized for photon energy resolution and had already proved itself at SPEAR by mapping out the energy spectrum of the three photon lines from ψ' to χ_c transitions in charmonium. As soon as the Crystal Ball came into operation at DORIS, the running concentrated on the $\Upsilon(2S)$. The Crystal Ball produced the best looking data on the $\Upsilon(2S) \rightarrow \gamma\chi_b(1P)$ lines, but ARGUS, CLEO and CUSB had decent data, too.

Later, after a run on the $\Upsilon(1S)$, the Crystal Ball group claimed a significant peak in the high energy inclusive photon spectrum, which they named the zeta (ζ), suggesting that it might be the Higgs boson. This prompted DORIS and CESR both to make long runs at the $\Upsilon(1S)$ energy. No one saw any sign of the ζ, and the Crystal Ball had to retract their discovery. After that the Crystal Ball retired from the field, leaving ARGUS as the only high energy physics experiment at DORIS. The Crystal Ball never attempted the radiative spectrum from the $\Upsilon(3S)$; CUSB had the best data on that.

In 1981 CLEO and CUSB made an energy scan of the total cross-section above the $\Upsilon(4S)$. Peaks appeared at 10.865 and 11.019 GeV center of mass energy. We assumed they were the next two $^3S_1$ $b\bar{b}$ states and we named them $\Upsilon(5S)$ and $\Upsilon(6S)$. At the time, they offered no advantages over $\Upsilon(4S)$, so we did not spend much time there.

When not running at the CUSB choice of energy, or chasing the zeta, or bump hunting at high energies CLEO ran at the $\Upsilon(4S)$ and concentrated on B meson decay physics. Because of the high B mass, the typical decay produces a large multiplicity of particles and the number of different exclusive decay channels is huge. Kinematic reconstruction of B decays was therefore very difficult. Most of our data came from inclusive rate measurements. That is, we would measure the total rate for some particle species at the $\Upsilon(4S)$ resonance energy, and subtract the corresponding rate at an energy just below $B\bar{B}$ threshold (called "the continuum"), scaled to account for the E^{-2} dependence of the continuum cross-section. Since the $\Upsilon(4S)$ was assumed to decay entirely to B^+B^- and $B^0\bar{B^0}$ approximately equally, the subtracted inclusive rates corresponded to the decay rates averaged over the two B charge states.

The standard running mode settled down to two weeks of resonance alternating with one week of continuum. We measured and published papers on inclusive B decays to e, μ, $K^\pm$, K_S, D^0, D^*, D_s, ϕ, ψ, baryons, charmed baryons, and all charm (see Appendix, Tables 11 and 12). We measured charged multiplicities in B decays, and from various combinations of inclusive lepton rates we obtained upper limits for neutrinoless leptonic decays, flavor changing neutral currents, $b \to u$ transitions and $B^0\bar{B^0}$ mixing. So far, no surprises. The b quark decay was consistent with charged-current decay to the next lighter doublet; that is, $b \to cW^-$, no flavor changing neutral currents, no exotic modes, and little or no $b \to uW^-$. The b quark was acting like the charge $-1/3$, lower-mass member of a weak doublet, and the six-quark Standard Model seemed to be working well. There had to be a top quark and we expected it would be found momentarily at PETRA, then at TRISTAN, then at LEP. We assumed that there was a window in time for b quark physics; as soon as the top quark was found, upsilons and B's would no longer be interesting. As it eventually turned out, the 174 GeV mass of the t-quark is so high that it does not form bound meson states before it decays, and there is essentially only one decay mode, $t \to bW^+$. Except for its possible coupling to a Higgs boson, the study of the top quark is not likely to provide much new information about the Standard Model or its generalizations.

As the B event sample got larger CLEO started to look for reconstructable exclusive decay modes. In this sense "reconstructing" a decay exclusively means identifying all of the daughter particles in the decay and verifying that energy and momentum conservation works back to the unique parent mass. The only modes for which the number of spurious random mass combinations would be low were those with the lowest final state multiplicities. And since the favored decay path was $b \rightarrow c$, there would be a D in the final state, for which we also required a low multiplicity decay. The typical product branching ratio we were looking for was therefore only $\sim (5 \times 10^{-3})(4 \times 10^{-2}) = 2 \times 10^{-4}$, for $B^- \rightarrow D^0\pi^-$ followed by $D^0 \rightarrow K^-\pi^+$, for instance. Sheldon Stone [12], at that time a Cornell Research Associate, led a group that found the first evidence of a mass peak, and in January 1983 CLEO submitted a *Phys. Rev. Lett.* [**50**, 881 (1983)] showing a signal in four modes combined: $D^0\pi^-$, $D^0\pi^+\pi^-$, $D^{*+}\pi^-$ and $D^{*+}\pi^-\pi^-$, with B branching ratios 4.2, 13, 2.6 and 4.8%, respectively, with large statistical errors. Better measurements a few years later from ARGUS and CLEO showed that our early measurements were mostly wrong. The reported B mass was two standard deviations (5 MeV) too low and the branching ratios were an order of magnitude too high. Most likely, we had indeed seen B decays, but we did not realize how many modes with an additional unseen pion could feed down and masquerade as simpler modes with only a minimal effect on the reconstructed mass. This was a rare example of CLEO publishing a wrong result, outside of quoted error limits.

In 1986, CLEO submitted to *Phys. Lett.* B [**183**, 429 (1987)] the results of its first (unsuccessful) search for rare exclusive B decay modes via the effective neutral current $b \rightarrow sg$ and $b \rightarrow s\gamma$ loop mechanism, called penguin decays because of the alleged resemblance of the Feynman diagram to a penguin. Although the branching ratios are of the order of 10^{-5}, the efficiencies are high because a secondary D meson need not be constructed. The upper limits we reported were all above 2×10^{-4}, though, and not yet in the interesting range.

At energies below the $B\bar{B}$ threshold 4/10 of the hadronic annihilation cross-section is charm production, $e^+e^- \rightarrow c\bar{c}$, and even at the $\Upsilon(4S)$ resonance there is more charm production than $b\bar{b}$. As we learned to reconstruct D's and D^*'s in B decays, it became obvious that CLEO had significant capability for charm physics. Although the final states in $e^+e^- \rightarrow c\bar{c}$ were more complicated than those at the SPEAR threshold energies, they were no worse than B decay final states, and the higher CLEO energies actually brought some advantages. We could be sure that any charmed particles found in the continuum data runs, or with momenta above the ~ 2.5 GeV/c maximum for B decay secondaries, had to be from direct $e^+e^- \rightarrow c\bar{c}$ production and not from $B\bar{B}$ events. In the CLEO-1 era 11 charm papers were published: inclusive cross-sections, momentum spectra, and decay branching ratios for D^*, D_s, Λ_c, Σ_c and Ξ_c, as well as D lifetimes and a search for charm changing neutral currents (see Appendix, Tables 8–10).

The most notable of the early CLEO charm papers was the "discovery" of the F meson, now called the D_s. Sheldon Stone and others found a clear mass peak at 1970 ± 7 MeV in $\phi\pi$ consistent with the decay of a $c\bar{s}$ meson. The problem was that DASP had in 1979 reported a few $\eta\pi$ events at a different mass, 2030 ± 60 MeV, apparently confirmed by a CERN OMEGA photoproduction experiment which had claimed to see events in $\eta\pi$, $\eta\pi\pi\pi$, $\eta'\pi\pi\pi$ and $\phi\rho$ at masses centered on 2030 ± 15 MeV. Moreover, the Mark I detector at SPEAR had run above the $e^+e^- \to D_s\overline{D_s}$ threshold and had not reported seeing anything. We were sure our data were right, however, so we prepared it for publication. When our Albany collaborator, Saj Alam, saw the draft he revealed that when he had been a member of the Mark I collaboration he had worked on the F search and had convinced himself that they were seeing $F \to \phi\pi$ at a mass around 1970 MeV, but the statistical weight of their meager data sample was not enough to convince his colleagues. Sheldon called Burt Richter to see if Mark I wanted to be referenced in our paper, and was told that they preferred that we not mention their earlier unpublished work. Our paper, "Evidence for the F meson at 1970 MeV" A. Chen *et al.*, *Phys. Rev. Lett.* [**51**, 634 (1983)], was greeted with scepticism for a while, but was confirmed by TASSO and ARGUS at DESY, and by ACCMOR at CERN. Whether you should count the D_s as a bona fide CLEO discovery depends on whether you think the earlier F claims were spurious background fluctuations or mass mismeasurements. A Fermilab proposal to look for the tau neutrino, based on the expectation of $D_s \to \tau\bar{\nu}_\tau$ events in a beam dump, was withdrawn because our lower mass for the D_s significantly reduced the expected branching ratio to $\tau\bar{\nu}_\tau$.

Beyond upsilon, B, and charm physics there was a miscellany of topics that accounted for ten papers: four on tau leptons, two on inclusive particle production in the non-$B\bar{B}$ continuum, one on Bose–Einstein correlations, and one each on unsuccessful searches for axions, multipoles and the $\xi(2200)$ seen by Mark III.

Just as noteworthy as some of the papers that CLEO published was one that we did not publish. Some time after the finding of the D_s, Hassan Jawahery, a Research Associate with Syracuse, discovered a narrow mass peak in $K^+K^-\pi^\pm$ somewhat below the mass of the D meson. Although the statistical significance of the "Jawaherion" looked moderately compelling, more than three standard deviations as I recall, it did look suspicious to most collaboration members. There was no plausible explanation for the existence of such a narrow state at that mass, and there was no confirming evidence in other modes. For more than a year we suppressed the news until we could check it with more data. No one mentioned it in public, not even Jawahery, who to his credit respected the will of the majority. The later data showed no peak; the original signal was apparently only an unlikely statistical fluctuation in the background.

The last CLEO paper (submitted in late 1988) based entirely on data obtained with the CLEO-1 detector before the installation of the new DR2 drift chamber, dealt with Σ_c^{++} and Σ_c^0 charmed hyperon continuum production and decay. It listed 91 authors from 12 institutions (see Appendix, Table 2). LeMoyne College and Rutgers University had left the collaboration, but seven new university groups had joined. The new CLEO institutions were acquired in two ways: (a) a CLEO Research Associate takes a faculty job at another university and leads a new group that joins CLEO (Ohio State, Florida, Minnesota, Maryland); or (b) a university with no previous connection to CLEO petitions to join (Albany, Carnegie Mellon, Purdue).

In March 1983 McDaniel decided that it was time to reopen the question of what best use could be made of the north interaction region. The Columbia–Stony Brook (CUSB) collaboration had been running their nonmagnetic NaI and lead-glass detector there since the turn-on of CESR in 1979, and had done good work on upsilon spectroscopy. The CUSB collaboration had grown to include Richard Imlay and others from Louisiana State University and Eckart Lorenz and others from the Max Planck Institute in Munich. LSU had contributed a muon detector surrounding the rest of the CUSB detector and Munich had added small-angle sodium iodide scintillator arrays. McDaniel called for proposals for the experiment to replace CUSB. The Program Advisory Committee met in January 1984 to choose between two north area options:

- a Columbia–Stony Brook proposal to upgrade CUSB (to CUSB-II) by replacing the tracking chambers with a cylindrical array of bismuth germanate (BGO) scintillation counters;
- a proposal by the UPSTATE collaboration (Caltech, Carnegie Mellon, Louisiana State, MPI Munich, Princeton, and Stanford, with Donald Coyne and Eckart Lorenz as cospokesmen) for a new compact detector with tracking chambers inside a BGO ball inside a superconducting solenoid.

Upon the recommendation of the PAC, McDaniel decided for the CUSB-II proposal. This was when the Carnegie Mellon group decided to join CLEO.

9

Improving CESR, 1981–1988

•

In the early days of CESR operation the accelerator physics activity was concentrated on learning how to operate the linac, synchrotron, storage ring complex reliably. Maury Tigner's positron coalescing scheme worked, but its complexity made injection difficult. Eventually the linac was upgraded to allow positrons to be produced at a rate sufficient to do without coalescing.

Like all the other storage rings proposed in the 1970s, CESR was supposed to have a peak luminosity of $10^{32}/\text{cm}^2\text{sec}$ [6]. It was calculated from the formula

$$\mathcal{L} = \frac{N_{e+}N_{e-}f_c}{4\pi\sigma_x\sigma_y}$$

under the following parameter assumptions.

	5 GeV	8 GeV
Beam energy	5 GeV	8 GeV
Luminosity, $\text{cm}^{-2}\text{sec}^{-1}$	0.59×10^{32}	1×10^{32}
Number of bunches per beam	1	1
Number of particles per bunch, $N_{e\pm}$	1.34×10^{12}	1.5×10^{12}
Circulation frequency, f_c	390.134 kHz	390.134 kHz
Horizontal r.m.s. beam size, σ_x	1.0 mm	1.0 mm
Vertical r.m.s. beam size, σ_y	0.09 mm	0.06 mm
β_y^*	10 cm	10 cm
Vertical tune shift, ξ	0.110	0.061

Like all the other storage rings, CESR never achieved this luminosity under these conditions. One reason was that we never had good reason to run at the

8 GeV per beam design energy. But more importantly, the nonlinear focusing effects of the beam–beam interaction at high beam currents made it impossible to keep to the quoted beam size. The beam–beam effect is parametrized by the linear tune shift ξ. It was found empirically that as the beam current is increased ξ reaches a saturation value beyond which the beam size increases in such a way that the luminosity becomes proportional to N instead of N^2. The limiting tune shift value is not well defined, but tended to be about 0.03, instead of the much higher value implied by the CESR luminosity projection. Moreover, we were not able to collide beams with more than about 2×10^{11} particles per bunch, presumably because of the destabilizing wake field effects of the interaction between the beam and the vacuum chamber and rf cavity walls.

In the saturated tune shift limit the luminosity is given in the usual c.g.s. units by

$$\mathcal{L} = 2.17 \times 10^{32} \frac{EeNf\xi}{\beta_y^*},$$

where E is in GeV, e is the electron charge in coulombs, N is the number of particles per bunch, f is the frequency of bunch passages, and β_V^* is the focusing depth of field parameter in meters. With $N = 2 \times 10^{11}$ and $\xi = 0.03$ the formula implies a luminosity of only $4 \times 10^{30}/\text{cm}^2\text{sec}$ at 5 GeV. By the end of 1980 we had reached 3.1×10^{30}, and it was clear that we had to do something drastic to get much further.

If N and ξ are limited, the only remaining variables are f and β_y^*. The latter parameter is proportional to the focal length of the interaction region quadrupole doublet. Since all the interesting physics was at energies well below the 8 GeV design energy, it was relatively easy to move the quads in closer to the CLEO and CUSB experiments and increase the currents in them, once we had learned to do without the solenoids that compensated for the focusing effect of the CLEO solenoid. We found that the compensation could be done with rotated quadrupoles located outboard of the I.R. focusing quads. So in the summer of 1981, we did it and reduced β_y^* to 3 cm; this was called the "minibeta" configuration. By the end of the year the luminosity had reached 8×10^{30} and by the following year it was 1.2×10^{31}.

Increasing the bunch frequency f would not be so straightforward. That would require storing more than one bunch per beam. The beauty of the original single bunch scheme was that the electrons and positrons could travel in precisely the same orbits and would collide head-on at only two points. With n bunches per beam there would be $2n$ collision points, all but two occurring where there were no detectors and where larger β values would imply enhanced beam–beam disruption that would seriously limit the achievable luminosity at the CLEO and CUSB interaction points.

Prof. Raphael Littauer came up with the solution by proposing that we install electrostatic fields outboard of each of the two interaction regions. The fields would deflect the electrons and positrons oppositely, each trajectory making several horizontal betatron oscillations through half the ring before returning to the undeflected orbit just before the next interaction region. These "pretzel" orbits would separate the electrons and positrons transversely by several cm at the $2n - 2$ undesired interaction points. The maximum bunch number $n = 7$ would be limited by the betatron tune of the ring.

The electrostatic separators were built and installed in June 1983, but making the multibunch configuration work was not easy. The separators had to maintain very high fields over several meters length, and every time there was a spark or high voltage glitch, the stored beam would be lost. Injection into pretzel orbits was difficult, and the available aperture in the beam chamber for keeping the beams separated from each other and from the walls was barely enough for usable beam lifetimes. For a while CESR ran with $n = 3$ bunches per beam, but the multibunch luminosity gain was only about ×1.5. The missing factor of 2 was apparently the effect of the aperture limitations. By 1985 we had reached a peak luminosity of 3.9×10^{31}. Prof. David Rubin succeeded in improving integrated luminosity in the same year by working out a procedure for keeping the stored positrons at the end of a beam run instead of dumping them; positron injection was much faster in this "topping-up" scheme.

The next step was "microbeta", a further reduction in β_y^* to 1.5 cm. In order to take the minibeta idea a step further we had to make quadrupoles that would fit inside the hole in the CLEO magnet pole and come within 60 cm of the center of the detector. Research Associate Steve Herb built permanent magnet quadrupoles to be supported from the end plates of the CLEO drift chamber. They were installed in 1986.

Since the β function goes through an hour-glass minimum at the interaction point, low β_y^* can help only over an interaction region length $\sigma_z/\sqrt{2}$ that is of the order of β_y^*. The bunch length σ_L is determined by the rf overvoltage and was about 2 cm. By 1988 CESR was running with luminosity 10^{32}, a record for e^+e^- rings. The improvement came not only from microbeta, but from increasing the number of bunches per beam to the limit of $n = 7$. The full benefit of microbeta, however, would have to wait for more rf voltage.

The rf system, in fact, was becoming a major headache. The original intent when CESR was designed had been to run at 8 GeV per beam with two 14-cell rf cavity assemblies. At 5.3 GeV only one of them was required to make up the 1 MeV/turn energy loss due to synchrotron radiation. But the rf cavities were severely stressed by the increasing power transferred to the beam as the number of bunches increased, and by the higher voltages required for bunch shortening. We had chronic problems with arcing at the input power windows and with vacuum leaks at many of the welds. Of the three cavity assemblies we had on hand, usually only one was in working order. It was frustrating to diagnose and fix problems. A bad cavity had to be removed from the ring, the water jacket had to be drained and dried out, the problem had to be located and fixed, the cavity had to be baked and pumped down, then it had to be vacuum tested and power tested. If the tests failed, as they did often, the procedure would have to be repeated. The turnaround time could be several months. The cavities were becoming the major reason for lost running time. Eventually, Don Hartill solved the window arcing problem, but the vacuum troubles continued to get worse.

10

The First Upgrade, CLEO-1.5, 1984–1989

———————————— • ————————————

In 1981 the remaining CLEO detector components were completed and installed (see Appendix, Table 1): the superconducting solenoid coil, the rest of the muon detection chambers, and all eight octants of dE/dx measuring chambers. Dave Andrews had designed the coil more conservatively than the Berkeley TPC coil he had copied it from, and although it was supposed to run at a field of 1.2 tesla, we never had the courage to run it above 1.0. As a result, it ran well and never had the kind of mishaps that the TPC coil had.

In 1984 a ten-layer "VD" cylindrical drift chamber built at Ohio State replaced the original IZ proportional chamber. Although the inner radius was a conservative 8.1 cm, we had trouble initially with synchrotron radiation and had to learn to be careful to mask out stray radiation with absorbers near the beam line. The improved spatial resolution close to the beam pipe made it possible to reconstruct separated decay vertices for charmed mesons and taus.

Although the CLEO detector was an excellent match to the requirements of the early upsilon and B meson physics at CESR, as time went on and the easy measurements had all been done, we became more aware of the limitations of the detector:

- the arrangement of cells in the drift chamber was such that at particular ϕ angles the sense wires lined up radially making it impossible to resolve the left–right drift ambiguity for tracks in those directions;
- although the solenoid coil was only 0.7 radiation length thick, a sizable fraction of the particles passing through it interacted, thereby degrading the information coming from the outer detector components, particularly the DX proportional chambers and the RS shower chambers;

- the poor $17\%/\sqrt{E_{\text{GeV}}}$ energy resolution of the RS proportional tube and lead calorimeter made it pretty useless for looking at photons from upsilon radiative transitions, or for reconstructing π^0's and η's in B decays;
- the MU chambers were so far away from the interaction point that decay muons from pions and kaons were a significant background;
- the detector was fragmented into too many separate subsystems, causing the less useful ones (OZ, CS, etc.) to be ignored.

We began to realize that the ARGUS detector being built at DESY might out-perform CLEO.

It was clear that we could make a significant improvement if we could accomplish the shower detection and the K/π separation inside the solenoid coil, but it was not immediately clear how. Informal groups within CLEO started working on both problems. We discussed three options for particle identification: (a) a drift chamber with more layers than the present one, optimized for dE/dx measurements as well as tracking, (b) a time projection chamber with dE/dx measuring capability as in the TPC experiment at PEP, and (c) a conventional drift chamber supplemented with a ring imaging Cerenkov detector. After heated debate we chose option (a) as being simpler and less risky, although not optimally effective for hadron identification at high momentum. The shower calorimeter discussions quickly centered on a high-Z scintillator: sodium iodide, cesium iodide, bismuth germanate or barium fluoride. Cesium iodide, although not widely used, seemed to offer a good compromise of performance and price, so we decided to expose some prototypes in a SLAC test beam.

To get started, we fixed the new DR2 drift chamber outer radius at 91 cm, the same size as the existing DR1 drift chamber. This would allow us to build the new one and run it in the original CLEO-1 detector, then later install it in a new CLEO-2 detector. There would probably be enough money in the annual operating budgets to build DR2, but we would have to convince the NSF to support a sizable capital upgrade project to build the rest of CLEO-2; that is, the new time-of-flight scintillators, CsI scintillator array, superconducting coil, magnet iron and muon detector.

The Cornell team that built the new drift chamber included David Cassel, Gil Gilchriese (who left soon afterward for the SSC project), Dan Peterson and Riccardo DeSalvo, a very energetic and inventive Research Associate from Pisa and CERN. The chamber was a close-packed array of single-sense-wire square cells arranged in 51 cylindrical layers. Through most of the chamber the sequence was $000+000-\ldots$, where 0 is axial, and $+$ and $-$ are slanted at angles of a few degrees plus and minus to give stereo information. The sense wire positions in successive axial layers were staggered by half a cell, to help resolve the left–right ambiguity in drift azimuth. Instead of an inner and outer layer of field wires, there were layers of cathode hoop strips to give extra measurements of the z coordinate. There were a total of 12,240

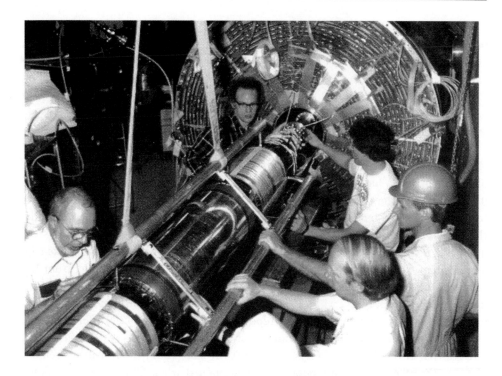

Fig. 18. Installation of the microbeta quads and the VD drift chamber into the new DR2 drift chamber. Clockwise from left are Joe Kirchgessner, Steve Herb, Mike Ogg, Bryan Kain and Steve Gray.

gold-plated tungsten 20 μm sense wires and 36,240 110 μm field wires, most of them gold-plated aluminum — a very large number of wires to string.

The completed DR2 drift chamber replaced the original DR1 during a five month shutdown in 1986 (see Fig. 18). The microbeta quadrupoles, installed at the same time, required a reduction in the beam pipe radius from about 7.5 cm to 5 cm, so we filled the space between the new beam pipe and the VD with a new 3-layer straw tube drift chamber called VD-insert or IV, built by Ohio State. Along with the tracking chambers we also replaced one of the two proportional-tube+lead end-cap shower detectors with a prototype cesium iodide array, to get running experience with the new kind of calorimetry. We ran this "CLEO-1.5" detector configuration for the next two years while we were building the new CLEO-2 magnet and outer detector components. The increase in number of drift chamber tracking layers from the original 17 of DR1 to the 64 of IV+VD+DR2 improved the pattern recognition and momentum resolution, and the reoptimized electronics allowed us to get 6.5% resolution in dE/dx from up to 61 pulse heights on a track. I rewrote the original SOLO track finding program to take advantage of the half-cell stagger in triplets of axial layers, hence the new program name, TRIO.

The modified detector came into useful operation quickly, but it took a while to build up enough of a data sample to rival that from the previous six years. In all, 33 CLEO-1.5 papers were submitted for publication between October 1988 and January 1992 (see Appendix, Tables 5–12). Six were on upsilon spectroscopy, 5 on B semileptonic decays, 8 on B hadronic decays, 8 on charmed mesons, 3 on charmed baryons and 3 on the tau. The highlights were the measurements of $B\bar{B}$ mixing and of the b-to-u decay

The first measurements of the B meson lifetime, made at PEP in 1983, had yielded an unexpectedly high value, about 1.2 picoseconds. Since the neutral B, like the neutral K, had no conserved quantum number that could distinguish it from its antiparticle, the slow decay rate raised the possibility that B^0 might oscillate to $\overline{B^0}$ or vice versa before it decayed. When the neutral B decays semileptonically, the sign of the lepton can label its flavor at decay time, so one could look for $B^0\overline{B^0}$ mixing to $B^0 B^0$ or $\overline{B^0}\,\overline{B^0}$ by running at the $\Upsilon(4S)$ resonance and observing like-sign lepton pairs with momenta high enough to exclude the cascade leptons from the semileptonic decays of D's from B decays. In 1986, CLEO reported an upper limit for the mixing probability: $\chi_d < 19\%$. Meanwhile, the UA1 CERN $\bar{p}p$ collaboration had seen like-sign dimuon events and reported an average $\bar{\chi} = 12 \pm 5\%$ for the B_d, B_s and Λ_b mixture (mostly B_d) produced at high energies. This was confirmed by ARGUS [H. Albrecht *et al.*, *Phys. Lett.* **B192**, 245 (1987)] with a measurement of $\chi_d = 17 \pm 6\%$. We were scooped! This happened when CLEO was shut down for the DR2 installation. ARGUS had accumulated a data set comparable to the CLEO data set, and their superior shower detector and shorter path for background from $\pi \to \mu\bar{\nu}_\mu$ gave them more dilepton sensitivity. This was their finest hour. In February 1989 we had enough data from CLEO-1.5 to make a high statistics remeasurement; we reported $\chi_d = 16 \pm 6\%$, confirming the ARGUS result.

In lowest order the b quark was expected to decay to a lighter, charge$= 2/3$ quark through the charged current weak interaction, that is, by emission of a virtual W^-. Whether it would be b-to-c or b-to-u would depend on the values of the off-diagonal Cabibbo–Kobayashi–Maskawa (CKM) matrix elements V_{cb} and V_{ub}. The CLEO inclusive B decay data were consistent with b-to-c dominance, and everyone wanted to know whether V_{ub} was zero or not. The Kobayashi–Maskawa mechanism for CP violation depended on V_{ub} being nonzero. There were two main ways to look for b-to-u. One could look for hadronic B decay modes that did not have a charmed particle or charmonium in the final state, for example, $B^0 \to \pi^+\pi^-$; so far, from CLEO-1 we had only upper limits [P. Avery *et al.*, *Phys. Lett.* **B183**, 429 (1987)]. Or one could look at the lepton momentum spectrum in semileptonic B decays, $\bar{B} \to X_u \ell^- \bar{\nu}$, beyond the end-point for $\bar{B} \to D\ell^- \bar{\nu}$. A lower mass recoil, say $\pi\ell^-\bar{\nu}$ for instance, would allow a higher momentum lepton. Since semileptonic decays were the best understood theoretically, this second method had the advantage that a signal could be more reliably related to the value of the V_{ub} CKM matrix element. CLEO-1 data had not shown a high momentum excess [S. Behrends *et al.*, *Phys. Rev. Lett.* **59**, 407 (1987)], implying an upper limit $|V_{ub}/V_{cb}| < 0.16$.

In mid 1988 we got word from ARGUS that they had seen charmless B decays into $p\bar{p}\pi$ and $p\bar{p}\pi\pi$, that is, evidence for the b-to-u transitions. These were modes we had not looked for. Had we been scooped again? There were not yet enough CLEO-1.5 data to check the ARGUS claim, so we went back to the CLEO-1 data. They were not conclusive. I had to give a review talk on nonleptonic B decays at the Stanford Heavy Flavor Symposium in July. I waffled. "CLEO confirms the effect in the $p\bar{p}\pi$ channel, but not in the $p\bar{p}\pi\pi$. As to whether it is really from correctly reconstructed charmless B decays, or from some yet to be determined cocktail of spurious misreconstructions, ...one has to start out with a skeptical bias ...I am suspicious of the biases inherent in the back-to-back angle cut. Lastly, the $\sin^2\theta$ distribution of the CLEO candidates, while not conclusive, does not favor the ARGUS hypothesis."

By October we had accumulated 212 pb^{-1} of CLEO-1.5 data, which when combined with the 78 pb^{-1} of older CLEO-1 data, was sufficient to confront the ARGUS report, based on 103 pb^{-1}. With more data, the indication that we had seen in $p\bar{p}\pi$ had gone away, and in the first paper based on CLEO-1.5 data [C. Bebek *et al.*, *Phys. Rev. Lett.* **62**, 8 (1989)] we were able to set upper limits well below the branching ratios claimed by ARGUS. We had not been scooped; the ARGUS results were spurious. Some CLEO members claimed that ARGUS had tuned their event selection cuts to pump up a statistical fluctuation. To avoid this bias, cuts must always be determined *a priori*, without reference to the actual data, say by using Monte Carlo simulated events.

We returned to the lepton momentum spectrum, and a year later had convinced ourselves that we had a high momentum excess that could come only from $B \rightarrow X\ell^-\bar{\nu}$ with recoil hadron masses m_X below that of the lightest charmed meson [R. Fulton *et al.*, *Phys. Rev. Lett.* **64**, 16 (1990)]. ARGUS confirmed it almost immediately, but we had scooped them. Indeed, $|V_{ub}|$ is not zero, it is 5 to 10% of $|V_{cb}|$. This discovery established the main outlines of the KM matrix governing quark weak decays and provided a basis for the violation of CP invariance. Suddenly B meson physics had caught fire. Here was a chance to understand the mystery of the baryon–antibaryon abundance asymmetry in the universe.

In 1988 we ran for a couple of months (113 pb^{-1}) on the $\Upsilon(5S)$. My student, Sumita Nandi, and a Kansas student, Sangryul Ro, looked in the data for evidence of the B_s ($=\bar{b}s$) meson. There were several marginal indications: the inclusive lepton momentum was best fit with a $B_s/(B+B_s)$ fraction $f > 41\%$; the inclusive D_s rate was consistent with $f = 30 \pm 18\%$; the dearth of reconstructed B decays implied $1 - f < 59 \pm 34\%$; and five exclusive B_s hadronic decay candidates suggested $f > 10\%$. The problems were (a) the low cross-section at the $\Upsilon(5S)$, (b) the relatively high background from the u, d, s, c continuum, (c) the unknown background from B and B^* production, and (d) the fact that contributions from $B_s\overline{B_s}$, $B_s\overline{B_s^*}$, and $B_s^*\overline{B_s^*}$ caused three different mass peaks for reconstructed B_s. CLEO concluded that the evidence was not good enough to claim the discovery of the B_s. From time to time, the collaboration debated taking more $\Upsilon(5S)$ data, but always decided that LEP or the Tevatron collider could produce B_s much better than CESR. Eventually, several LEP experiments did publish convincing evidence for the B_s and also the Λ_b.

There was one paper that CLEO did publish, but later wished it had not. In the course of our study of inclusive ψ production in the non-$B\bar{B}$ continuum we took a look at the $\Upsilon(4S)$ also. We had known for some time that there were ψ's from the sequence,

$$e^+e^- \rightarrow \Upsilon(4S) \rightarrow B\bar{B} \rightarrow \psi XX',$$

with momenta below the kinematic limit, $(M_B^2 - m_\psi^2)/2M_B = 2.0$ GeV/c. We were surprised, however, to see a signal for higher momentum ψ's, at a rate above that from the background measured below $B\bar{B}$ threshold, that is, $0.22\pm0.07\%$ per $\Upsilon(4S)$. Apparently, the $\Upsilon(4S)$ had a significant branching fraction to non-$B\bar{B}$ final states. This was completely unexpected, since the decay rates of the lower upsilon states (to non-$B\bar{B}$, of course) were $\Gamma < 0.05$ MeV, compared with $\Gamma(\Upsilon(4S) \rightarrow B\bar{B}) = 24$ MeV. Later we discovered that we had been fooled by an unlikely statistical fluctuation in the non-$B\bar{B}$ continuum under the $\Upsilon(4S)$. The probability of a fluctuation beyond three standard deviations is only 0.26%, but if you make enough measurements, you will eventually be stung.

This is an appropriate point at which to say more about two of the stars of the CLEO collaboration. Although there are many members whose contributions have been outstanding, Sheldon Stone and Ed Thorndike are especially noteworthy. Sheldon started out as a junior faculty member at Vanderbilt when CLEO was first organized and joined the Cornell group as a Research Associate in the early data taking period. After a while as a Cornell Adjunct Professor he became a Full Professor with the Syracuse group. Sheldon always has strong opinions on what CLEO should be doing. Although his aggressive advocacy sometimes annoys his colleagues, he is almost always right. Whenever CLEO is in a building or upgrading phase, Sheldon picks the most demanding project and leads the effort to a successful state-of-the art detector subsystem. The CLEO-2 cesium iodide detector and the CLEO-3 ring imaging Cherenkov system are two examples that I will detail in later chapters. In data analysis, Sheldon goes for high profile, important topics. As I have already noted, he led the reconstruction of the first exclusive B decays and discovery of the D_s.

Ed Thorndike of the University of Rochester faculty has impeccable physics judgment and an instinct for CLEO discovery opportunities. He manages to attract excellent grad students and post-docs, and with their help, he searches for new physics in the CLEO data, often using novel techniques of his own invention. He and his team are responsible for many limits on forbidden processes and first observations of highly suppressed processes. The most famous of his discoveries is probably the so-called "penguin" decay of the B's, that I will describe in due course.

Ed is a skilled manager, especially in situations of contentious decision making and time-sensitive building projects. He has served as elected CLEO spokesman more than anyone else — seven years.

11

CLEO-2, CESR and CHESS Upgrades, 1985–1989

•

A detailed progress report [16] on the CLEO-2 design and prototype work (CBX-83/77) was presented to the Program Advisory Committee in December 1983; more definitive updates (CLNS 84/609 and 85/634 [15]) appeared in May 1984 and January 1985. Major NSF funding started in November 1984, and continued for five years. The total upgrade cost was $37,380,000, of which $23.2 million went for the CLEO-2 detector and the rest for upgrading CESR and the Laboratory computing facility.

It is hard for me now to explain why the NSF was willing to support such a costly upgrade effort. We were asking to spend about $14 million on cesium iodide alone. Our competition at the time was the Laser Interferometry Gravitational Observatory proposal (LIGO). Some possible reasons for our success are (a) the CLEO-2 design was elegant and superbly adapted for the physics goals, (b) the CLEO collaboration had a good reputation for cost effective detector design and construction, and for productive exploitation of the physics potential, (c) the competition from ARGUS gave a sense of urgency, (d) we were ready to go and LIGO was not, (e) the CHESS program in X-ray science would benefit from the planned increase in CESR circulating currents, and (f) we had the confidence and enthusiastic support of David Berley, our NSF Program Officer, and of Marcel Bardon, the Physics Division Director. Maybe they wanted to cheer us up after our defeat on the CESR-II proposal.

McDaniel appointed Bernie Gittelman to manage the CLEO upgrade project (see Figs. 19, 20, and 22). I list below the major components and the institutions primarily responsible for them. Other groups also had responsibilities for various parts of the electronics, software, and so on.

Fig. 19. (top) The CLEO-2 magnet during installation. (bottom) Technician wiring the preamps on the CLEO-2 cesium iodide scintillator array.

Fig. 20.   The CLEO-2 detector in the L-0 pit, with the west pole being inserted.

PT	inner 6-layer straw tube drift chamber	Ohio State (Kagan)
VD	intermediate 10-layer drift chamber	[existing]
DR2	main 51-layer drift chamber	Cornell (Cassel)
TF	barrel scintillator trigger and TOF array	Harvard (Pipkin)
CC	cesium iodide shower scintillator array	Cornell (Stone)
	superconducting solenoid coil and iron	Cornell (Nordberg)
MU	muon proportional chambers	Syracuse (Moneti)
TFend	end-cap TOF scintillator array	Albany (Alam)
CCend	end-cap cesium iodide array	Cornell (Kubota)
LM	small angle luminosity monitor	Carnegie Mellon (Engler)

The most expensive and time consuming part of the CLEO-2 construction effort was the cesium iodide scintillator array (see Fig. 19, bottom). This was a new kind of detector on a scale never before attempted. Many new problems had to be solved:

- specifying the cesium iodide purity, doping and surface quality to be sufficient for good acceptable energy resolution while keeping the crystal cost as low as possible;
- finding a way to pay the vendors (BDH in England and Horiba in Japan) enough up front to set up for mass production, without breaking the first year's budget;
- getting crystals delivered at a rate consistent with installation of the full detector in 1988.

- finding photodiodes and preamplifiers that would have acceptably low noise and low cost;
- mounting the 7800 barrel scintillators in a robust, almost-massless structure with each crystal aimed at the interaction point;
- guaranteeing a failure rate low enough to survive many years of no access to the detector.

The details of the various solutions are given in several instrumentation publications (Appendix, Table 13).

Oxford Instruments in England made the superconducting coil. Compared with the CLEO-1 solenoid, it was about 50% larger, ran at 50% higher field, and was much thicker, since now only the muons had to get through it to be detected. Most of the magnet iron (see Fig. 19, top) was machined out of pieces of the SREL synchrocyclotron magnet (Newport News, VA) by Dominion Bridge in Montreal. The Syracuse group built the muon detection chambers in the Iarocci style, that is, plastic channels with one anode wire per channel and crossed cathode strips. Instead of the Iarocci streamer mode, they used proportional mode, since the electronics was already available from the CLEO-1 dE/dx system.

The last few weeks before the installation shutdown Riccardo DeSalvo made a test of a 2 cm radius insert in the interaction region beam pipe to see how close we could get a detector to the beam without being overwhelmed by backgrounds. The test was marginally successful; we backed off and decided on a 3.5 cm radius. Ohio State built a 5-layer straw tube PT (precision tracker) replacement for the former IV chamber. The installation shutdown started in April 1988 and lasted until August 1989. The CLEO-1 magnet was shipped off to Brookhaven.

In parallel with the hardware effort, Cornell post-doc Rohit Namjoshi led a team to write the software required for the new detector. They decided to throw out the rather cumbersome CLEO-1 structure that had evolved over the past decade and start fresh with a system built on the ZEBRA data-base program from CERN. For a while they seemed to be stuck in the block-diagramming stage, but they eventually emerged with some new code. Nobu Katayama was largely responsible for the data base management. The tracking was still based on TRIO for online and DUET for offline. There were two cesium iodide photon algorithms, one from Brian Heltsley, and the other derived by Tomasz Skwarnicki who had worked with the Crystal Ball experiment. The Monte Carlo simulation of the detector was based on the CERN GEANT program. All in all, the system worked quite well, although later, as we got more experience and got more clever with corrections that improved the efficiency and resolution, each data set got recompressed (reduced to data summary files) at least twice.

During the shutdown for the installation of CLEO-2 there were improvements made also to CESR and to CHESS. The CESR improvements, aimed at higher luminosity, were mostly evolutionary upgrades designed to make multibunch and microbeta work better. We began to realize as the upgrade progressed that the main obstacle to higher luminosity was the rf system. Running with two 14-cell cavities (Fig. 10) was definitely an improvement over running with one — it shortened the bunches, making the microbeta more effective — but it was getting increasingly difficult to keep two of the three 14-cell cavities in working order. As the CLEO upgrade was nearing completion well within the projected budget, the money was available to replace all the cavities with new ones. The NSF approved the reprogramming of upgrade funds, so we resolved to build four 5-cell cavities, plus one more for a spare. The new cavities had the following advantages:

- they were tuned for higher beam currents;
- we could put more power into each cell with 5 cells per rf window instead of 14;
- more of the vacuum joints were electron beam welded;
- the outer water jacket was made more easily demountable;
- the high order mode probes were more accessible;
- we corrected other mistakes in the original design.

The rf upgrade program started as the rest of the upgrade effort was coming to an end; the last of the 14-cell cavities was finally retired in 1993.

In order to keep CHESS competitive with existing and planned synchrotron radiation facilities at Brookhaven, Berkeley, Argonne, and other places, Boris Batterman and I felt that we had to expand the number of X-ray beam lines available. After toying for a while with the idea of digging into the hillside west of the existing CHESS areas to create more space, we realized that it would be more economical to build a CHESS-East area to use the X-rays emitted by the positron beam on the opposite side of the south interaction region from the existing CHESS(-West) electron beam area. This CHESS upgrade not only doubled the CHESS experimental facilities, but provided the opportunity to create a special station dedicated to irradiation of biologically hazardous specimens at the BL-3 level. The new CHESS-East area also got a 24-pole wiggler magnet to create very high intensities.

12

The CLEO-2 Years, 1989–1995

•

Following the recommendation of the Program Advisory Committee, we spent the first six months or so after the CLEO-2 installation running at the $\Upsilon(3S)$ energy. CUSB had the priority, and CLEO used the data to tune up the new detector. The monoenergetic photon lines from radiative transitions between $b\bar{b}$ bound states were especially useful for calibrating the cesium iodide. Although the first few months of data were rather ragged, the CLEO-2 detector turned out to be a great success, exceeding its projected performance goals in every respect. Never before had there been a detector with simultaneous momentum–energy resolution for energetic charged particles and photons better than 2%. CLEO published four papers based on the $\Upsilon(3S)$ data (see Appendix, Table 5). The first one showed the power of the cesium iodide calorimeter; the three photon lines from $\Upsilon(3S) \rightarrow \chi_b(2P)\gamma$ were beautifully resolved. In another paper we photon-tagged the $\chi_b(2P_0)$ and $\chi_b(2P_2)$ decays to gg in order to make a direct comparison between gluon jets and quark jets from continuum $q\bar{q}$ production.

Following the $\Upsilon(3S)$ run CESR did an energy scan in the region of the $B\overline{B^*}$ threshold to investigate the first excited B meson state, called B^*. In the first published paper based on CLEO-2 data, (see Fig. 21 and Appendix, Table 12) CLEO used the inclusive rate for the 46.2 MeV photon line from $B^* \rightarrow B\gamma$ to measure the $B^* - B$ mass difference (improving on the earlier CUSB data) and the energy dependence of the inclusive B^* production cross-section. The experiment was motivated by Sheldon Stone's suggestion that one could produce $B\bar{B}$ in a charge conjugation +1 state via $e^+e^- \rightarrow B\overline{B^*} \rightarrow B\bar{B}\gamma$, and thus measure CP violation in $B^0/\overline{B^0} \rightarrow \psi K$ interfering with mixing, without having to observe the time dependence, as you would have to do in the favored scenario with $C = -1$ pairs from $e^+e^- \rightarrow B\bar{B}$. This would have allowed CESR to measure CP violation

VOLUME 67, NUMBER 13   PHYSICAL REVIEW LETTERS   23 SEPTEMBER 1991

Measurement of the Inclusive B^* Cross Section above the $\Upsilon(4S)$

D. S. Akerib,[1] B. Barish,[1] D. F. Cowen,[1] G. Eigen,[1] R. Stroynowski,[1] J. Urheim,[1] A. J. Weinstein,[1] R. J. Morrison,[2] D. Schmidt,[2] M. Procario,[3] D. R. Johnson,[4] K. Lingel,[4] P. Rankin,[4] J. G. Smith,[4] J. Alexander,[5] C. Bebek,[5] K. Berkelman,[5] D. Besson,[5] T. E. Browder,[5] D. G. Cassel,[5] E. Cheu,[5] D. M. Coffman,[5] P. S. Drell,[5] R. Ehrlich,[5] R. S. Galik,[5] M. Garcia-Sciveres,[5] B. Geiser,[5] B. Gittelman,[5] S. W. Gray,[5] D. L. Hartill,[5] B. K. Heltsley,[5] K. Honscheid,[5] J. Kandaswamy,[5] N. Katayama,[5] D. L. Kreinick,[5] J. D. Lewis,[5] G. S. Ludwig,[5] J. Masui,[5] J. Mevissen,[5] N. B. Mistry,[5] S. Nandi,[5] C. R. Ng,[5] E. Nordberg,[5] C. O'Grady,[5] J. R. Patterson,[5] D. Peterson,[5] M. Pisharody,[5] D. Riley,[5] M. Sapper,[5] M. Selen,[5] H. Worden,[5] M. Worris,[5] P. Avery,[6] A. Freyberger,[6] J. Rodriquez,[6] J. Yelton,[6] K. Kinoshita,[7] F. Pipkin,[7] R. Wilson,[7] J. Wolinski,[7] D. Xiao,[7] A. J. Sadoff,[8] R. Ammar,[9] P. Baringer,[9] D. Coppage,[9] R. Davis,[9] P. Haas,[9] M. Kelly,[9] N. Kwak,[9] H. Lam,[9] S. Ro,[9] Y. Kubota,[10] J. K. Nelson,[10] D. Perticone,[10] R. Poling,[10] S. Schrenk,[10] M. S. Alam,[11] I. J. Kim,[11] B. Nemati,[11] V. Romero,[11] C. R. Sun,[11] P.-N. Wang,[11] M. M. Zoeller,[11] G. Crawford,[12] R. Fulton,[12] K. K. Gan,[12] T. Jensen,[12] H. Kagan,[12] R. Kass,[12] R. Malchow,[12] F. Morrow,[12] J. Whitmore,[12] P. Wilson,[12] F. Butler,[13] X. Fu,[13] G. Kalbfleisch,[13] M. Lambrecht,[13] P. Skubic,[13] J. Snow,[13] P.-L. Wang,[13] D. Bortoletto,[14] D. N. Brown,[14] J. Dominick,[14] R. L. McIlwain,[14] D. H. Miller,[14] M. Modesitt,[14] E. I. Shibata,[14] S. F. Schaffner,[14] I. P. J. Shipsey,[14] M. Battle,[15] H. Kroha,[15] K. Sparks,[15] E. H. Thorndike,[15] C.-H. Wang,[15] M. Artuso,[16] M. Goldberg,[16] T. Haupt,[16] N. Horwitz,[16] V. Jain,[16] R. Kennett,[16] G. C. Moneti,[16] Y. Rozen,[16] P. Rubin,[16] T. Skwarnicki,[16] S. Stone,[16] M. Thusalidas,[16] W.-M. Yao,[16] G. Zhu,[16] A. V. Barnes,[17] J. Bartelt,[17] S. E. Csorna,[17] T. Letson,[17] and M. D. Mestayer[17]

(.EO II Collaboration)

f Technology, Pasadena, California 91125
Santa Barbara, Santa Barbara, California 93106
iiversity, Pittsburgh, Pennsylvania, 15213
'orado, Boulder, Colorado 80309-0390
iversity, Ithaca, New York 14853
Florida, Gainesville, Florida 32611
iity, Cambridge, Massachusetts 02138
ollege, Ithaca, New York 14850
f Kansas, Lawrence, Kansas 66045
nnesota, Minneapolis, Minnesota 55455
w York at Albany, Albany, New York 12222
University, Columbus, Ohio 43210
)klahoma, Norman, Oklahoma 73019
rsity, West Lafayette, Indiana 47907
ochester, Rochester, New York 14627
iversity, Syracuse, New York 13244
iiversity, Nashville, Tennessee 37235
eceived 20 May 1991)

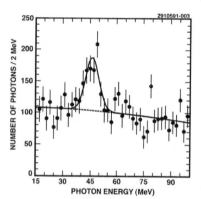

Using the CLEO II detector at the Cornell Electron Storage Ring, we have determined the inclusive B^* cross section above the $\Upsilon(4S)$ resonance in the energy range from 10.61 to 10.70 GeV. We also report a new measurement of the energy of the $B^* \to B\gamma$ transition photon of $46.2 \pm 0.3 \pm 0.8$ MeV.

PACS numbers: 13.65.+i, 11.30.Er, 13.40.Hq, 14.40.Jz

We report a measurement of the inclusive B^* cross section as a function of center-of-mass energy just above the $\Upsilon(4S)$ resonance. The value of the $B\bar{B}^*$ [1] cross section is necessary to determine the feasibility of observing time-integrated CP-violating asymmetries at a symmetric B factory [2]. We can also compare the hadronic cross section in this energy region with predictions of several potential models [3]. The inclusive B^* cross section is determined by measuring the yield of photons from the transition $B^* \to B\gamma$. The branching fraction for this transition is 100% because the mass difference between the B^* and the B mesons is too small to allow the emission of a pion.

The data used in this analysis consist of 57.8 pb^{-1}

Fig. 21.   First paper based on CLEO-2 data.

Fig. 22.   The time-of-flight scintillators in the CLEO-2 detector.

with equal beam energies, but unfortunately the $B\overline{B}^*$ production rate turned out to be seven times lower than the $B\bar{B}$ rate at the $\Upsilon(4S)$. From then on, practically all of the running was at the $\Upsilon(4S)$ resonance and immediately below $B\bar{B}$ threshold.

Accelerator physicists had made a convincing presentation to the PAC [16] that more beam luminosity could be obtained for the south interaction point (CLEO) if CESR were operated without collisions in the north (CUSB). Upon the advice of the PAC, I decided to terminate the CUSB experiment after one more run on the $\Upsilon(4S)$. The new CLEO-2 cesium iodide calorimeter could do everything that the

CUSB BGO-plus-NaI detector could, and had good resolution for charged particles as well. CUSB under the leadership of the Franzinis had run in the north interaction region for eleven years and had accomplished a lot of good physics with a small collaboration, with limited resources, under hardship conditions. Among other accomplishments, they discovered the radiative transitions to the $\chi_b(1P)$ and $\chi_b(2P)$ states in the upsilon system, and they obtained the first indication for the B^*. The CUSB detector was dismantled, the trailers on Upper Alumni Field were hauled away, and the CUSB experimenters went on to work with the D∅ experiment at Fermilab. Later the Franzinis moved to Frascati to set up the KLOE experiment at the DAΦNI phi factory collider. The luminosity advantage for CLEO was in fact real; by the end of 1990 it had reached $1.5 \times 10^{32}/\text{cm}^2\text{s}$.

The CLEO collaboration had begun to grow more rapidly during the construction of the new detector. The first CLEO-2 paper had 133 authors from 17 institutions. The new institutions that had joined in the CLEO-1.5 era (see Appendix, Table 2), Kansas, Oklahoma, UC Santa Barbara, Colorado and Caltech, represented a westward shift in the center of gravity of the collaboration and included a group previously in ARGUS (Kansas) and groups from the SLAC orbit (the latter three). The new groups also broadened the spectrum of CLEO physics interests; for example, UC Santa Barbara (Morrison, Witherell, *et al.*) brought experience in charm decays from the Fermilab E-691 experiment, and Caltech (Barish, Stroynowski, *et al.*) was especially interested in tau physics. As the collaboration grew, Cornell became less dominant, and the proportion of DOE supported groups increased to about two-thirds.

With increasing size the collaboration became more bureaucratic (see Appendix, Table 3). In 1990 the CLEO Analysis Coordinator, David Besson, set up a number of Physics Topic Analysis groups (PTA's): B semileptonic decays, hadronic B decays, rare B decays, charmed mesons, charmed baryons, taus and strong interaction physics (a miscellany of upsilon spectroscopy, two-photon physics, fragmentation, etc.). The monthly CLEO meetings were supplemented the preceding or following day by PTA meetings, where most of the physics discussions took place. It became impossible for one individual to keep track of all the physics analysis activities going on in the collaboration. Some members in fact were interacting only with their PTA's and were unaware of anything else. I guess this sort of trend is inevitable in large organizations. The average shift running obligation per CLEO member was getting so sparse (shifts were manned by two physicists) that it was difficult to maintain continuity and familiarity with the running of the experiment.

There was not enough space in Wilson Lab to accommodate the increasing number of CLEO collaborators (see Appendix, Table 2) and transient CHESS users. The extra wing added on the west side of the lab in 1985 was already inadequate. Several times LNS and CHESS submitted to the NSF a proposal for adding a fourth and fifth floor to the lab. It failed, either because we were not able to get Cornell to commit matching money, or because the guidelines for NSF infrastructure grants excluded additions to existing facilities. So we set up modular units (like mobile homes) in the yard, three in 1989, five in 1993 and more in 2001. Space was a perennial problem.

The increased CLEO manpower and the new ability to reconstruct kinematically the decay final states with π^0's and γ's along with charged particles boosted CLEO's physics productivity. Up through year 2000, 167 CLEO-2 papers were submitted for publication (see Appendix Tables): 18 on semileptonic B decays, 39 on nonleptonic B decays, 27 on charmed mesons, 19 on charmed baryons, 34 on τ leptons, 11 on upsilon bound states, 5 on two-photon processes and 4 others. CLEO had developed a well-oiled machinery for producing papers. A small group or even a single individual would circulate a draft and make a presentation at a meeting. The collaboration would vote on whether it was publishable, perhaps in amended form, and if so, in which journal. The Analysis Coordinator would appoint a committee of experts to meet with the authors to suggest more work, rewrite the paper or whatever they felt necesssary. The new version would be circulated to the collaboration with an invitation for comments. If the Analysis Coordinator felt that the changes were warranted, there be another collaboration vote before submitting for publication.

The CLEO-2 efficiency and resolution for observing π^0 and η led to measurements of previously inaccessible τ branching ratios: $h^- n\pi^0 \nu_\tau$ ($h = \pi$ or K, $n = 1, 2, 3, 4$), $\pi^- \pi^+ \pi^- \nu_\tau$ and $\pi^- \pi^0 \eta \nu_\tau$. All the single-charged-prong tau decays were measured and previous measurements of the major tau branching fractions were improved, with the result that Martin Perl's claim of a deficit of exclusively determined decay rates relative to inclusive sums disappeared. Tau decay branching ratios agreed well with Standard Model predictions; no surprises there.

In the charmed meson domain CLEO made definitive measurements of the branching ratios for the five D^* to D transitions, involving $\pi^\pm$, π^0 and γ. Previous SPEAR measurements had been quite wrong. Using D's tagged by the D^* to D transition, CLEO made precise absolute measurements of the branching ratios for the normalizing D decay modes to $K^-\pi^+$ and $K^-\pi^+\pi^+$, important for charm cross-section determinations. The spectroscopy of the $L = 1$ D^{**} states was explored, and CLEO's results on the semileptonic decays of the D tested the newly revealed heavy quark effective theory. The list of measured branching ratios was extended to include good measurements of $\pi^+\pi^-$, doubly Cabibbo suppressed $K + \pi^-$ and decays involving $\overline{K^0}$ and $\overline{K^{*0}}$.

CLEO also continued its dominance of the physics of the strange charmed mesons. The spectroscopy was advanced by papers on the $D_{s1}(2536)$, the $D_{s2}^*(2573)$ (a new discovery), and the $D_s^* - D_s$ mass difference. New D_s decay modes involving η or η', plus π or ρ were studied. A measurement of the semileptonic decay $D_s \rightarrow \phi \ell \nu$ put the absolute normalization of all the D_s modes on a firmer basis and produced more form factor measurements for heavy quark effective theory. The highlight of the D_s work was the observation of the purely leptonic $D_s^+ \rightarrow \mu^+ \nu_\mu$

decay. Since this has to proceed through the annihilation of the c and $\bar{s}$ quarks, it provides a measure of the decay constant f_{D_S} characterizing the quark–antiquark bound state overlap. The interpretation of $B^0 - \overline{B^0}$ mixing in terms of Cabibbo–Kobayashi–Maskawa matrix elements depends on knowledge of the decay constant for heavy-quark + light-antiquark, and the leptonic D_s decay is so far the best source of experimental information.

Collaborators from Albany, Florida, Carnegie Mellon and Ohio State specialized in charmed baryons. They produced a wealth of discoveries that made CLEO the prime source of data on baryon states containing the c quark. Several papers were published on the decay modes of the Λ_c — $pK\eta$, $\Lambda\pi$, $\Lambda\eta\pi$, $\Lambda K\bar{K}$, $\Lambda\ell\nu$, $\Sigma n\pi$, $\Sigma\eta$, $\Sigma\rho$, $\Sigma\omega$, $\Sigma K\bar{K}$, $\Sigma^*\eta$, ΞK, $\Xi K\pi$; the $\Lambda_c^*(2593)$ and Σ_c^+ were discovered; and the decays $\Xi_c \to \Omega K$ and $\Xi_c \to \Xi\ell\nu$ were observed.

During this period there was talk at several laboratories about building a Tau–Charm Factory, a high luminosity e^+e^- collider to operate in the $c\bar{c}$ threshold region. For a while people at SLAC, Seville, CERN, Dubna, Novosibirsk and Beijing were enthusiastic, but eventually financial realities and the wealth of tau and charm data coming out of CESR and LEP discouraged almost all of them. Martin Perl and the rest of the SLAC tau-charm partisans eventually joined the CLEO collaboration. By 2001 it appeared that funding would be obtained for a Beijing tau–charm factory and that CESR would be modified to run at charm threshhold. Many of the original tau–charm goals have in the meantime been achieved at CESR operating at $b\bar{b}$ threshold energies and by LEP experiments at higher energies.

Although the close proximity of the CLEO detector and the interaction region beam focusing quadrupoles makes electron tagging of the two-photon process, $e^+e^- \to e^+e^- hadrons$, impossible in CLEO, the very high effective rates for $\gamma^*\gamma^* \to hadrons$ makes it tempting to do $\gamma^*\gamma^*$ experiments in which only the *hadrons* are detected. The UC San Diego group, which had formerly been part of the PEP TPC-Two-Photon collaboration, joined CLEO in 1991 with the idea of exploiting the detector for two-photon physics. Previous measurements elsewhere of $\gamma^*\gamma^* \to \pi^+\pi^-$, K^+K^- and $p\bar{p}$ were carried to higher energies and $\gamma^*\gamma^* \to \chi_{c2}$ was observed.

Of course, the *raison d'être* for CLEO-2 was B physics. With a better detector and a larger data set we could improve many of the measurements that had been made with CLEO-1 and CLEO-1.5 — the semileptonic branching ratio, exclusive B to charm decay modes, mixing and V_{ub} — and also push down the limits on various b-to-u modes, semileptonic and those involving a D_s, as well as $B \to \ell^+\ell^-$. Several interesting new decays were observed: $B \to \Sigma_c X$ and the Cabibbo- and color-suppressed $B \to \psi\pi$.

Most notable was the discovery of several rare charmless decay modes. $\overline{B^0} \to \pi^+\pi^-$ involves V_{ub} by having the b and its partner $\bar{d}$ exchange a W to become $u+\bar{d}$, and $\overline{B^0} \to K^-\pi^+$ is an effective neutral current b-to-s transition that can occur through an intermediate $W^- + c$ (or u or t) state. The Feynman loop diagram, complete with the spectator $\bar{d}$ and emitted gluon, was alleged by John Ellis to resemble a penguin in order to win a bet (or repay a debt) by getting "penguin" printed in *Phys. Rev. Lett.* CLEO observed a peak in the reconstructed beam constrained B mass, consistent with a branching ratio of $(2.4\pm0.8) \times 10^{-5}$ to either $\pi^+\pi^-$ or $K^-\pi^+$ or a mixture of both. In order to suppress the rather serious non-$B\bar{B}$ background, a Fisher discriminant variable was formed from an optimal linear combination of several variables that had marginally different distributions for signal and background. The optimal combination was found by Monte Carlo simulation. There were two ways to separate the $\pi\pi$ and $K\pi$ hypotheses: energy conservation ($E_h + E_\pi = E_{\text{beam}}$), and dE/dx measurements on the tracks. Each method gives somewhat less than two standard deviations separation between π and K, so the best we could do for the individual modes with the available statistics was to quote upper limits. This measurement was either the first observation of an exclusive b-to-u mode or the first observation of a hadronic penguin decay, or both. It caused quite a stir among the theorists. Both channels are important for the measurement of CP violation in B decays.

Along with the $b \rightarrow sg$ hadronic penguin modes, one expects also $b \rightarrow s\gamma$ radiative penguin modes, the most likely channel being $K^*\gamma$. A Rochester–Syracuse–Cornell team spearheaded by Ed Thorndike (see Fig. 23) found signals in three charge combinations: $K^-\pi^+\gamma$ (Fig. 24), $K^-\pi^0\gamma$ and $K_S^0\pi^-\gamma$. The challenge in this measurement was the suppression of background from non-$B\bar{B}$ continuum $q\bar{q}$ jet events. Several distributions that were expected by Monte Carlo studies to be slightly different for signal and background were used to form a likelihood ratio for each event, for which a selection cut was defined. Standard Model expectations are rather explicit for inclusive $b \rightarrow s\gamma$, but there is no theoretical consensus on how much of the s quark fragmentation should show up as $K^*(890)$. So although the measured average branching ratio, $\mathcal{B}(B^{+,0} \rightarrow K^{*+,0}\gamma) = (4.5 \pm 1.7) \times 10^{-5}$, was the first confirmation of the existence of the penguin mechanism, it was not a quantitative test of the theory.

This was remedied soon by an inclusive measurement of $b \rightarrow s\gamma$. The problem again was background suppression, but in this case, we were looking not for a reconstructed B mass peak, but only for a high energy photon, unfortunately not a distinctive signature. The extraction of the signal was a real tour de force. Two competing techniques gave consistent results. In the first, Ed Thorndike and his student, Jesse Ernst, combined eight event topology distributions into a single variable, using a neural net algorithm trained with Monte Carlo $b \rightarrow s\gamma$ signal and with background from continuum γ radiation and π^0 production. A fit of the data spectrum in this variable to Monte Carlo signal plus background spectrum shapes showed a significant signal. In the other analysis, Tomasz Skwarnicki required that the event have a reasonable χ^2 for reconstructing as a $K + n\pi + \gamma$ ($n = 1$ to 4). The idea was not that the reconstruction had to be literally correct, but only that requiring it would reduce background. The branching ratio averaged from the two

Fig. 23. Clockwise from upper right: Mike Billing, Steve Playfer (Syracuse), Peter Kim, Ed Thorndike (Rochester), Dave Rice and Yoram Rozen (Syracuse).

efficiency corrected analyses, $(2.3 \pm 0.7) \times 10^{-4}$, was consistent with the Standard Model prediction, using the value of $|V_{ts}|$ inferred from the measurements of b-to-u decays and $B\bar{B}$ mixing. Alternatively, it could be used for an independent measurement of $|V_{ts}|$. The amplitude for the loop diagram is sensitive to the presence

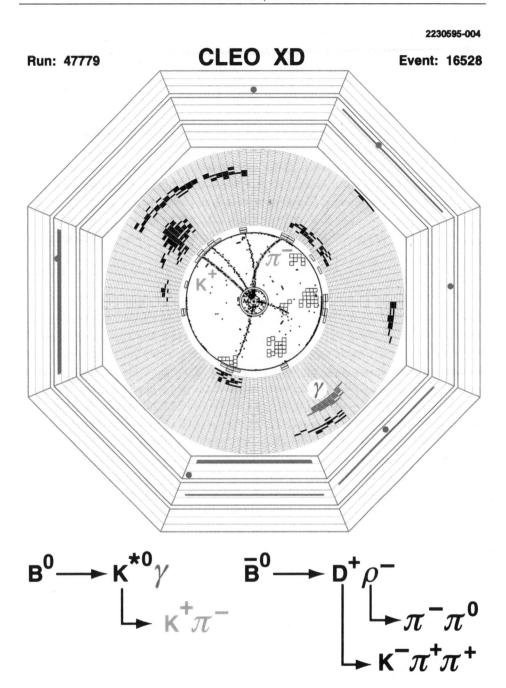

Fig. 24. End view computer display of a $e^+e^- \to B^0\overline{B^0}$ event in CLEO-2, showing a radiative penguin decay: $B^0 \to K^{*0}\gamma_1$, $K^{*0} \to K^+\pi_1^-$. The other tracks come from $\overline{B^0} \to D^+\rho^-$, $D^+ \to K^-\pi^+\pi^+$, $\rho^- \to \pi_2^-\pi^0, \pi^0 \to \gamma_2\gamma_3$ (γ_3 escapes unseen). The CsI shower counter array is shown with the near end at the outer circle.

of hypothetical particles. The agreement with the Standard Model could, for instance, be used to exclude a charged Higgs with a mass below 244 GeV, a much more restrictive limit than available by any other technique. The 1994 Lab holiday season card showed a snow scene with penguins.

Several important lessons followed from these first observations of rare B decays.

- The loop decays opened an exciting window on non-Standard Model physics, competitive with experiments at multi-TeV facilities.
- Although CESR had reached the luminosity level required to see the most prominent CKM-suppressed and penguin-loop decays, much more luminosity would be needed before CLEO could really explore the new field.
- CLEO was able to pioneer powerful, novel techniques for separating rare signals from copious backgrounds: cell list, Fisher discriminant, likelihood ratio, neural net.
- To capitalize on the potential of rare decays, CLEO would have to do a better job of $K - \pi$ separation at momenta above 1 GeV/c.

Semileptonic B decay (that is, with an e or μ in the final state) was a hot topic (see Table 8). Not only is the lepton a good flavor tag (i.e. picking out $b\bar{b}$ from everything else, or distinguishing b from $\bar{b}$ with the lepton charge sign), but the measurement of semileptonic branching ratios is our best source of information on the values of the CKM matrix elements V_{cb} and V_{ub}. The fact that the weak interaction is always $b \to W^- c$ (or $W^- u$) and $W^- \to \ell^- \bar{\nu}$ minimizes the confusion from multiple amplitudes, final state interactions, higher order effects, and so on; and the end point of the lepton momentum spectrum distinguishes $b \to W^- u$ from $b \to W^- c$. Comparison of the various exclusive rates as functions of the $\ell + \nu$ four-momentum q^2 tests the Heavy Quark Effective Theory used to understand the effect on the weak decays of the QCD heavy-light $Q\bar{q}$ dynamics. There were two discrepancies between theory and experiment. Naive counting rules corrected for phase space implied that the branching ratio $\mathcal{B}_{sl}$ for $B \to e\ell\nu$ should be about 16.5%. Corrections for hadronic enhancement of the nonleptonic rate could bring $\mathcal{B}_{sl}$ down as low as 11.5%. The CLEO inclusive data, however, implied $\mathcal{B}_{sl} = 10.4 \pm 0.3$. For a while it was possible to blame the disagreement on the model dependence of the correction for the unseen lower end of the experimental lepton momentum spectrum, but in the later measurements made with tagged events this was no longer a significant correction. Also, the LEP data, which initially implied a higher branching ratio, eventually fell into line with the CLEO result. The other problem was the fact that the sum of the measured exclusive branching fractions and upper limits for $B \to [D, \ D^*, \ D_1(2420), \ D_2^*(2460), \ \dots]\ell\nu$ accounted for only two-thirds of $\mathcal{B}_{sl}$.

In the CLEO-1.5 era the main competitor for CLEO had been the ARGUS experiment at DORIS. Over the years, the friendly rivalry had benefited both groups by raising the level of enthusiasm within and outside the two collaborations, and by keeping both sides honest through checking of each other's results. The prospect of a new and better CLEO-2 detector prompted the ARGUS experimenters to make their own upgrade, that is, a precision vertex drift chamber located next to the beam pipe. Rather than cylindrical, it was planar, with pentagonal symmetry, and was built by their Canadian collaborators. Unfortunately, it developed a short circuit after it was installed and much of the solid angle was lost. Later it was replaced by a silicon microstrip detector. This time it was an accidental beam overexposure during machine studies that ruined the detector. Ultimately, it was the fact that DORIS lost the luminosity race that was the undoing of ARGUS. Since the construction of PETRA in the late 1970s, DORIS never got the primary attention of the DESY accelerator physicists, even though for the 1980s, DORIS was the prime source of physics data for the laboratory. During the construction of HERA, DORIS luminosity continued to suffer from low priority, and half of the running time was dedicated to synchrotron radiation users. By 1993 the CESR peak luminosity had reached almost $3 \times 10^{32}/\text{cm}^2\text{sec}$, while DORIS had an order of magnitude less, so the running of ARGUS was terminated; actually, they had had very little successful data collection in several years. For a while there was no more competition for CESR in the $b\bar{b}$ threshold region.

The $b\bar{b}$ production cross-section is even higher at the Z^0 resonance than at the $\Upsilon(4S)$. As LEP gradually accumulated more integrated luminosity, and the properties of the Z^0 itself became well established, many of the ALEPH, DELPHI, L3 and OPAL experimenters turned their attention to B physics. B physics at high energies suffers from several disadvantages — accompanying fragmentation particles, poorer mass resolution, no reliable non-$b\bar{b}$ subtraction — but there are compensating advantages — resolvable decay vertices, separation of the B and $\bar{B}$ into separate jets, and concurrent production of B, B_s, B_c, Λ_b and other b-hadrons. The LEP experimenters learned to cope with the disadvantages and made good use of the advantages to produce results on the various b-hadrons, their lifetimes and semileptonic branching ratios, as well as on $B^0 - \overline{B^0}$ and $B_s - \overline{B_s}$ oscillations. LEP luminosity was never high enough, however, to rival CESR for rare B decays. Although there was some overlap, B physics at LEP tended to be mostly complementary to the work at CESR. When the LHC pp collider is completed, the LHC-B experiment will be the focus of B physics at CERN.

The CDF experiment at the Tevatron $\bar{p}p$ collider enjoys a large B production cross-section, but is almost overwhelmed by backgrounds. CDF has been able to pick out good signals for decays involving muons, such as ψK_s and ψK^*, allowing them to measure B lifetimes and mixing, and they have the tightest upper limit on the forbidden $B \to \mu^+ \mu^-$. For years there was talk at Fermilab of a dedicated B physics collider detector. Although the Fermilab PAC has approved the B-TeV proposal, funding at this date is still in doubt.

13

The CESR B Factory Proposal, 1989–1993

•

The CP operation, which takes particles into their mirror-image antiparticles, is not a symmetry of the universe, at least in our immediate vicinity, since we see mostly protons, neutrons and electrons and hardly any antiprotons, antineutrons and positrons. It is a puzzle how the universe got to be this way. Sakharov suggested that the asymmetry must be related to the 1964 discovery by Fitch, Cronin, Christianson and Turlay that CP symmetry is violated in about 0.2% of the weak decays of neutral kaons.

In 1973, even before the discovery of the charmed quark, Kobayashi and Maskawa realized that in the Standard Model the 3×3 quark doublet rotation matrix connecting the energy and flavor eigenstates for six quarks could give rise to CP violation at the level observed in kaon decay. It requires that each of the nine CKM matrix elements V_{ij} be nonzero and at least one be complex. If so, then CP violation should occur also in B decays, with the magnitude of the effect being proportional to the area enclosed by the triangle in the complex plane defined by the unitarity relation,

$$V_{td}V_{tb}^* + V_{cd}V_{cb}^* + V_{ud}V_{ub}^* = 0 \,.$$

After the CLEO discovery of b-to-u decays there was enough experimental information on the matrix elements to conclude that the enclosed area was indeed nonzero.

Ikaros Bigi, Tony Sanda and others worked out the various ways in which CP violation in B decays could manifest itself experimentally. The interference term between two amplitudes involving different KM matrix elements but with same exclusive final state could contribute with the opposite signs to CP conjugate B and $\overline{B}$ partial decay rates. For example, there would be an asymmetry between the branching ratio for $B^+ \to K^+\pi^0$ (which can occur through a $V_{ub}V_{us}^*$ tree amplitude or a $V_{cb}V_{cs}^*$ loop amplitude) and the branching ratio for $B^- \to K^-\pi^0$.

There are two problems with this, one experimental and the other theoretical. The decay modes with the largest predicted asymmetries $\mathcal{A}$ have very low branching ratios $\mathcal{B}$ and vice versa. The size of the $B\overline{B}$ event sample required for a statistically significant $\mathcal{A}$ measurement is of the order of $1/\mathcal{B}\mathcal{A}^2$, with $\mathcal{B} \sim 10^{-5}$ and $\mathcal{A} \sim 10^{-2}$ for a favorable mode. The event sample would have to be several orders of magnitude larger than available in the 1990s. But once an effect were observed, its interpretation would be confused by the fact that the asymmetry is proportional not only to the imaginary component in the KM matrix but also to the sine of the strong interaction phase in the decay amplitude, and such phases are not yet reliably calculated.

Both difficulties are made easier by considering instead the interference between the two ways that a neutral B can decay to a nonflavor-specific final state, like ψK_S or $\pi^+\pi^-$. The B^0 can either decay directly, or it can first oscillate to a $\overline{B^0}$ before decaying to the same final state. Since the mixing probability is rather large, the interference term can be significant. Another advantage is that the measurable asymmetry depends only on the CKM matrix elements and not on strong interaction phases. Two experimental complications arise, however. First, you cannot tell whether you started with a B^0 or a $\overline{B^0}$ by observing the final state, as you could in the $K^+\pi^0$ versus $K^-\pi^0$ case. You have to rely on the fact that the B's are produced in opposite flavor pairs, and *tag* the decay by observing the partner decaying into a flavor-specific mode, like $X\ell^\pm\nu$. Second, the two B's oscillate coherently in such a way that for a $B^0\overline{B^0}$ pair produced in a $C = -1$ state (as from a virtual photon in e^+e^- annihilation) the net asymmetry in the tagged rates is always zero. However, provided your detector has vertex resolution finer than the mean decay length, you can observe an oscillating *time dependent* asymmetry.

This does not yet solve the problem, because for $B\overline{B}$ produced from an $\Upsilon(4S)$ at rest the mean decay length is only 30 μm, which is difficult (though probably not impossible) to resolve experimentally. In 1987 Pier Oddone of Berkeley suggested increasing the decay length by boosting the $\Upsilon(4S)$ rest frame in the lab; that is, by colliding electrons and positrons of different energies. The two beams would have to be stored in separate, intersecting rings and brought to a common focus. The idea of separate rings for electrons and positrons was not new; Tigner (in CBN 82-24) had considered it as early as 1982 as a way of increasing beam currents. But the concept of asymmetric energies had to be checked out with realistic interaction region optics designs and beam–beam interaction simulations. A lot of design activity, workshops and internal debate at many laboratories — Paul Scherrer Institute (Villigen, Switzerland), CERN, DESY, Novosibirsk, KEK, Caltech, SLAC, and Cornell — along with numerous workshops around the world, resulted in serious "B Factory" plans at KEK and SLAC. I recall that at some conference around this time Dave Berley, our NSF Program Officer, surprised me in the corridors by

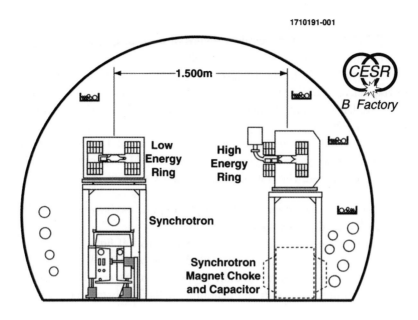

Fig. 25. Sketch of the tunnel cross-section showing the CESR-B rings and the synchrotron injector.

urging us to submit a Cornell proposal to the NSF to build an asymmetric B Factory. Although I favored the B Factory concept and I knew that we at Cornell had the capability, I protested that the cost was clearly off limits to the NSF Physics Division. I believe now that this was another example of the Physics Division "proposal pressure" strategy, that is, raising their profile in the competition with other divisions even when there was no hope of getting the high-cost proposals approved. Anyway, Maury Tigner and a team of enthusiasts put together a plan for a Cornell B Factory. Our four-volume CESR-B proposal [15] was submitted to the NSF, and the SLAC PEP-2 proposal was submitted to the DOE, simultaneously in February 1991.

The CESR-B luminosity goal was $3 \times 10^{33}/\text{cm}^2\text{sec}$ and the ring energies were 8 and 3.5 GeV. The design made use of the existing CESR 8 GeV ring and tunnel (see Fig. 25), the linac-synchrotron injector, and the CLEO-2 detector. The major items of new construction were a 3.5 GeV magnet ring, a copper vacuum chamber for both rings, a superconducting rf system, focusing magnets for the interaction region, and some additional building space. The CLEO-2 detector would be upgraded in data acquisition rate, vertexing capability, and particle identification.

In 1991 the agencies reviewed the CESR-B and PEP-2 proposals separately. An NSF cost review panel, chaired by Fermilab's Gerry Dugan, verified the CESR-B cost estimates and set the total project cost at $116 million, including upgrades to the CLEO detector. A DOE panel set the PEP-2 cost at $167 million *plus* the price of a new detector (unspecified, but about $50 million). It was pretty clear that at most only one of the two B Factory proposals would be funded, so the NSF and DOE decided to have a joint technical review of both proposals. But the NSF was already overcommitted on LIGO, and the DOE had no money for new initiatives beyond the Fermilab Main Injector project, so before the review could take place, William Happer (the DOE director of Energy Research) and David A. Sanchez (the NSF Associate Director for Mathematical and Physical Sciences) postponed further consideration indefinitely.

Meanwhile, with the encouragement and support of the NSF, under Maury Tigner's leadership we were carrying out a program of B Factory R&D. Maury had returned to Cornell from the SSC Central Design Group in 1989.

- We built a full size superconducting protoype rf cavity and tested it successfully to full field at the specified Q value.
- We tested a one-third size crab cavity model to full field.
- We tested a prototype high power window up to 250 kW.
- We measured the rf and vacuum properties of several brands of ferrite to confirm the practicality of the beamline loads for absorbing parasitic higher-order modes.
- CESR beam tests confirmed that there is no significant degradation in the luminosity for uncompensated beam crossing angles up to 2.8 mrad.
- Experiments with CESR tested ion trapping predictions.
- Theoretical studies and experimental measurements on transverse beam tails at high ξ were carried out.
- Titanum sublimation pumps installed for test in CESR resulted in significant improvement in the vacuum near the IR.
- We studied beam related backgrounds with a small beam pipe in CLEO-2.
- We tested the performance of a fast multibunch feedback electrode in CESR with high current bunches spaced by 28 ns.

Most of this work would be useful for the future performance of CESR whether or not a B Factory were built.

Although the DOE budget request for fiscal year 1994 did not originally mention funding for a B Factory, California lobbying efforts resulted in adding $36 million for the first year of construction of the SLAC B Factory. New York's Senator Daniel Patrick Moynihan, knowing that there was also a proposal from Cornell, insisted that there be a stipulation that the site be selected only after there was a review of both proposals. The eventual DOE budget request made public in April 1993 contained the wording, "In addition ... $36,000,000 ... provided that no funds may be obligated for construction of a B factory until completion, by October 31, 1993, of a technical review of the Cornell and Stanford Linear Accelerator proposals by the Department of Energy and the National Science Foundation." Suddenly and unexpectedly we were in direct competition with SLAC for DOE (not NSF) B Factory funding.

The DOE and NSF set up a review committee under the chairmanship of Stanley Kowalski of the MIT Bates Laboratory, and charged it to make a joint technical review of the CESR-B and PEP-2 proposals and to report back to the DOE Secretary by July 1993. Maury and his task force had about a month of hectic work to update our design and cost estimates and prepare our presentations. In June, the committee spent a week at SLAC and then a week at Cornell.

Each proposal involved supplementing an existing ring by building a new low energy ring, and the luminosity goals were the same, to be achieved in both cases by storing a large number of beam bunches. The CESR and PEP circumferences were 770 m and 2200 m, and the proposed energies were 8 & 3.5 and 9 & 3.1 GeV, respectively. In CESR-B the beams would collide at a 12 mrad crossing angle (necessitating rf transverse-mode "crab" cavities to compensate) and all rf cavities would be superconducting. In PEP-2 the beams would collide head-on and the rf cavities would be copper. Mainly because of the smaller circumference and the existing detector, the Cornell proposal would be about half the price of the SLAC proposal. One could characterize the Cornell proposal as taking maximum advantage of new technology, while the SLAC proposal was pushing old technology beyond where it was tested. There were technical risks in both, of course.

The Cornell review went very well, I thought, and we convinced the committee that we had a viable design and the ability to carry it out. The report supported our position, that both proposals could meet the goals of a B Factory and that the Cornell proposal would cost about $100 million less. Even so, it seemed to be too much to expect that the new DOE Secretary, Hazel O'Leary, would decide for Cornell in preference to a traditional DOE laboratory. Indeed, in October President Clinton announced on a trip to San Francisco that the B Factory was being awarded to SLAC. From the context of the President's announcement it seemed that the basis of the award was political and not technical.

This of course was a disappointment for us at Cornell, especially for those who had worked so hard on the CESR-B planning and R&D. There were some who were relieved, though, that we would not be getting involved with the DOE bureaucracy. Anyway, we picked up the pieces and managed to reconstruct a viable future program for CESR.

There was some consolation in the fact that the Japanese B Factory designers adopted many of the features of CESR-B, and that both the BaBar detector design for PEP-2 and the Belle detector design for KEK-B were patterned on CLEO-2.

14

CESR and CLEO Phase II Upgrade, 1990–1995

———————————— • ————————————

In the era of the 1990s there were nine operating high energy beam–beam colliders in the world. They are shown on the map in Fig. 26. Besides CESR there were five other electron–positron machines. In order of increasing energy they were the BEPC ring at Beijing operating down at charmed quark threshold, the VEPP-4 ring at Novosibirsk with a total energy of 14 GeV, the Tristan ring at the KEK lab in Japan at about 60 GeV, the SLAC Linear Collider (SLC) running at the 91 GeV Z resonance, and the CERN Large Electron–Positron ring (LEP) operating at and above the Z resonance energy.

By 1991 CESR had reached a luminosity plateau corresponding to peak values from 2 to 3×10^{32}/cm^{2}sec, and for the next four years it delivered between 1.1 and 1.5 fb^{-1} (inverse femtobarns) of integrated luminosity per year. We had squeezed all that we were going to get out of the microbeta, seven-bunch, one-interaction-region configuration. Although this was a world's record for luminosity, and we had beaten all the competition, it did not seem enough. When you run for more than four years at the same event rates, you eventually run out of interesting measurements that you can make in a reasonable time. Once we had discovered the first loop decay, $B \to K^*\gamma$ in 1993, we knew we had opened up a window on physics beyond the Standard Model, but at the rate we were going it would take forever to exploit it. The competition for the B Factory inspired accelerator physicists at Cornell to think creatively about luminosities like 3×10^{33}. Losing the B Factory would put CESR into competition eventually with facilities with that sort of performance.

The big question was how to put more beam bunches into CESR. When the beams collide head-on, the minimum longitudinal spacing Δs between successive bunches has to be longer than the 73 m spacing between the two electrostatic separators flanking CLEO, or else you get multiple collision points. Since this

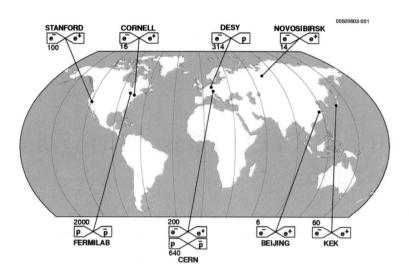

Fig. 26.   A map showing the location in the early 1990s of the world's high energy colliders. For each collider is indicated the particle species and their total energy in GeV.

distance is more than the length of one loop of the pretzel (typically $C/2Q_H = 768$ m$/(2 \times 9.4) = 41$ m), the multibunch scheme allows only as many bunches $n_{e+} + n_{e-}$ as can fit into separate loops in the pretzel; at the normal $Q_H \approx 9.4$ CESR betatron tune, that means $n = n_{e+} = n_{e-} = 7$.

One obvious idea was to increase the horizontal tune Q_H. For a while we considered a "CESR-plus" scheme to make more pretzel loops, raising Q_H to about 13.9, allowing $n = 14$ [Dave Rubin, CLNS 89/902 and CON 90-1]. But it turned out to involve major changes in the CESR lattice. The next idea was the "ΔE scheme", also invented by Dave Rubin. He suggested [CON 90-2] using a $\sim 1.8\%$ energy difference between the two beams, so that they could circulate at separate radii. By arranging the dispersion he could get the beam orbits to coincide at the experimental I.P. without using electrostatic separators. In order to have two equilibrium orbits at different momenta, you need to have different field integrals $\int B d\ell$ along the two orbits, so Dave envisioned a bypass in the north area for one of the beams. Since the two beams would avoid each other everywhere except near the south I.P., there would be no loop constraint and it would be possible to reduce the bunch spacing Δs to perhaps 24 m and thus store $n = 32$ bunches. But the effort and expense involved in the bypass gave us pause, and there was no adiabatic way of approaching the new configuration step by step. We would not know whether it would work until we had invested all the effort and money in building it.

In July 1990 Robert Meller wrote up a "Proposal for CESR Mini-B". Inspired by the B Factory proposal to intersect beams from separate rings at a $\alpha = 12$ mrad angle, he suggested using pretzel orbits in a single ring, crossing horizontally at $\alpha = 2$ mrad instead of colliding head-on as is usual in single ring machines

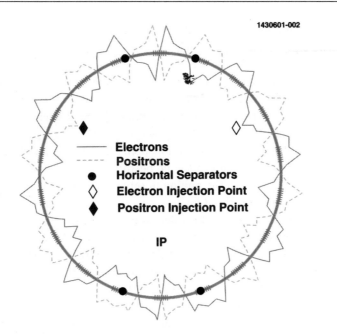

Fig. 27. Diagram of the separated pretzel orbits crossing at an angle at the interaction point. Distances transverse to the nominal orbit curve are exaggerated. The tick marks show the points where electron and positron bunches pass each other.

(α is the angle between the beam and the no-pretzel beam line). The fields in the electrostatic separators would be adjusted to continue the pretzel through the I.R. with a node at the center. This would permit filling each pretzel loop around the ring with a train of bunches spaced by a distance Δs just sufficient to separate electron and positron bunches by the minimum transverse miss distance at the first parasitic crossing point nearest to the desired intersection (see Fig. 27). The Meller scheme had several very attractive features.

- It would allow CESR to store up to five bunches in each pretzel loop, with a proportional gain in luminosity.
- It did not require extensive CESR lattice modifications.
- The various pieces of the scheme could be tried out one by one: asymmetric pretzels, crossing angles, bunch trains, high currents, and so on.
- Development of the technique would have the dual purpose of preparation for two-ring B Factory operation with high beam currents and nonzero crossing angle.

There were several problems to be solved.

1. The beam–beam interaction with an *angle crossing* can couple transverse and longitudinal oscillations and therefore excite synchro-betatron resonances.

This was the phenomenon that limited the luminosity of DORIS in the original two-ring configuration. Theory said that the effect would be minimal though if the crossing angle were less than the x-versus-z aspect ratio of the beam bunch: $\alpha < \sigma_x/\sigma_z$. This was verified in CESR; that is, the one-bunch-on-one-bunch luminosity was not significantly affected by crossing angles $|\alpha|$ up to 2.8 mrad with the 0.55 mm $\times$ 18 mm CESR bunches.

2. The long range beam–beam interaction at the *first parasitic crossing* a distance $\Delta s/2$ from the interaction point had to be minimized. The kick (horizontal or vertical) given to one beam bunch by the opposing one is proportional to $\sqrt{\beta_{H,V}}$, so the minimum separation Δs that you can achieve for a given α depends on the β functions at the parasitic crossing point. For the Δs values that we needed in order to make bunch trains of 2 or 5 bunches (Δs = 8.4 or 4.2 m), the parasitic crossings occurred near where β_H and especially β_V were going through large maxima in the final focus doublet. In order to get high luminosity we would have to reconfigure the interaction region quadrupoles for shorter focal lengths, bringing the β maxima in closer to the interaction point.

3. The crossing angle orbits made large excursions near the interaction point, coming closer to the *aperture limits* in the final focus quadrupoles, especially during injection. The quadrupole apertures would need to be enlarged.

4. The *long range beam–beam interactions* of the many electron–positron near misses all around the ring could give a cumulative tune shift effect that would limit the beam current and the luminosity, especially if the separate beam–beam kicks added coherently. Measurements were made under various conditions in CESR, and Sasha Temnykh invented an empirical model that seemed to fit most of the data. Extrapolated to the many-bunch, high-current regime where we had no good data, it suggested that the luminosity might increase only as the square root of the number of bunches instead of linearly.

5. Wakefield effects would be worse with more bunches and shorter Δs. To combat them Joe Rogers designed and installed a fast, digital multibunch *feedback* system to stabilize the beam. After some initial troubles, this was quite effective.

6. It was not clear how much beam current the four five-cell copper *rf cavities* could sustain. First, there was a limit to how much power one could couple through the cylindrical quartz windows in the input waveguides. But probably more importantly, the broad-band impedance of the cavities would allow the wakefields generated by the many high-current bunches to resonate and limit the achievable beam current. It is difficult to design a cavity shape with high shunt impedance (accelerating volts^2 per input watt) at the fundamental accelerating frequency and simultaneously low broadband impedance (wake volts per beam amp). The solution to this problem is superconducting cavities. With the low wall dissipation you can produce accelerating volts with

minimal input watts, thus giving you the option of sacrificing some shunt impedence in the cavity geometry to get low broadband impedence. The decreased power dissipation also allows you to put more power into the beam before reaching the window heating limit. Since the same reasoning holds for B Factory cavity design, the SRF group started developing a special single-cell superconducting cavity to be used either in a CESR B Factory or for the Meller-scheme luminosity upgrade.

7. The intense synchrotron radiation from high beam currents would bombard the *vacuum* chamber walls and cause increased heating and outgassing. Unless pumping speed were improved, the effect would be shorter beam lifetimes and increased backgrounds in the CLEO experiment. Step by step the most critical vacuum system components would have to be upgraded as the beam currents increase.

8. Increased beam currents require higher *injection* rates (amps per second). Fortunately, we were not trying to increase the per bunch linac currents, but upgrades would be necessary in the injection efficiency and in the power capability of the positron converter target.

9. The interaction region focusing could not be pushed significantly closer to the collision point (see item 2, above) without intruding on the *CLEO tracking chambers*, which would therefore have to be replaced.

After we had submitted the B Factory proposal in 1991, and received the Happer–Sanchez letter postponing indefinitely any action on the proposal, we began to make serious plans for the no-B Factory alternative, that is, a more modest CESR luminosity upgrade, combined with an upgrade of the CLEO detector. We decided on two stages, dubbed "phase II" and "phase III" by Dave Rice. Phase I corresponded to previous upgrades already completed.

Phase II included whatever we could do on a short time scale with minimal commitment of resources: (1) reconfiguration of the interaction region focusing with enlarged bore for the electromagnetic quads and lengthening the permanent magnet quads (using pieces from the ones retired from the north area), (2) new water cooled beryllium beam pipe and masking near the interaction point (see Fig. 28, top), (3) replacement of the CLEO PT straw tube chamber by a three-layer, double-sided, silicon detector (Fig. 28 bottom), (4) various CESR improvements to the vacuum, linac, and feedback system. Of the list of problems above, this would deal partially with #2, 3, 5, 7, 8, 9. The CLEO silicon tracker was motivated by (a) the hope that more charm decays would be identified, both from *B* decays and continuum production, and (b) the experience that we would get for the kind of measurements that would be important at a B Factory. Phase II would begin immediately, but anything beyond it could be displaced by CESR B Factory construction.

Phase III would involve more time, three or four years, and more money, over $30 million counting contributions from CLEO collaborators, and would therefore require a special NSF commitment in order to proceed. It would include (1) a

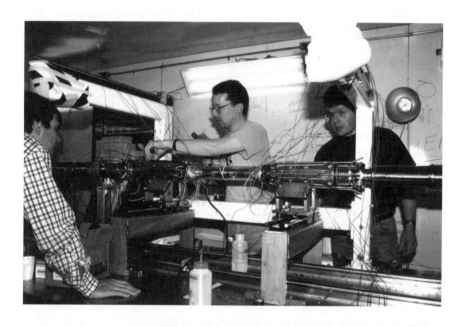

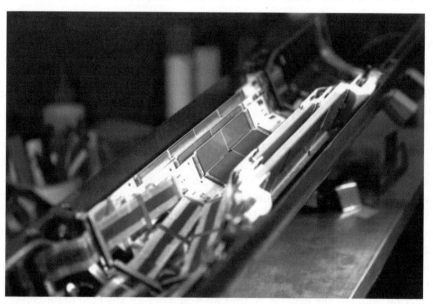

Fig. 28.   (top) Jeff Cherwinka, Denis Dumas and Ken Powers assembling the interaction region beam pipe for the phase II upgrade. (bottom) One half of the phase II CLEO silicon tracking detector.

superconducting upgrade of the interaction region quads, (2) superconducting rf cavities, (3) a rebuilding of the inner part of the CLEO detector, and (4) more improvements to the CESR vacuum and linac; addressing problems #2, 4, 6, 7, 8, 9.

One of the beauties of this plan was that it coincided almost exactly with the R&D to demonstrate the feasibility of the B Factory proposal, plus some parts of the actual B Factory construction. In early 1993 the phase III proposal was submitted to the NSF as part of the five-year proposal for CESR/CLEO operations for fiscal years 1994 through 1998. Meanwhile progress on phase II was being paced by the assembly of the silicon tracking detector for CLEO, which took longer than anticipated. The various CLEO responsibilities were apportioned as follows.

Beryllium beam pipe & masks	Harvard (Yamamoto)
Silicon detector assembly	UC Santa Barbara (Nelson)
electronics	Cornell, Illinois, ... (Alexander)
movable shielding	Cornell, Carleton (Dumas)
software	Cornell, UCSB, Ill. (Katayama)
VD repair and recabling	Ohio State (Kagan)
Pipe, silicon, VD assembly	Purdue, Cornell (Fast)
I.R. installation	Cornell (Kandaswamy)

The CLEO delays prompted the accelerator physics crew to run CESR in the crossing-angle, bunch-train mode before the installation of the interaction region focus modifications, even though the configuration was far from optimum for high luminosity. It went much better than anyone expected, and running in the 9-train × 2-bunch mode with $\alpha = 2.0$ mrad and $\Delta s/c = 28$ ns was declared the standard in early 1995. A month before the phase II installation shutdown began in April, CESR made a new luminosity record, $\mathcal{L}_{pk} = 3.2 \times 10^{32}/\text{cm}^2\text{sec}$.

The work during the shutdown involved more than just CLEO and the interaction region. The linac, vacuum system, electrostatic separators, and rf system were all refurbished, and the shielding between CESR and CHESS was upgraded. CLEO took advantage of the shutdown to repair a broken wire in the VD and to convert the gas system to operate the DR and MU chambers with a helium based gas instead of the former argon–ethane mixture. This involved the installation of a system for flushing nitrogen through the time-of-flight photomultiplier housings in order to avoid leakage of accumulated helium through the glass. There was a scare when the second half of the CLEO silicon detector (Fig. 28, bottom) from Santa Barbara came damaged, but the loss in number of good data channels turned out to be minimal. CESR and "CLEO-2.5" were turned on in October 1995.

15

The CLEO-2.5 Years, 1995–1999

—————————————— • ——————————————

The upgraded CLEO detector, called CLEO-2.5 or CLEO II.V, was a new device as far as charged particle tracking was concerned. Not only was the inner straw tube drift chamber replaced with three layers of double-sided silicon strips, but the gas in the main drift chamber was also changed from argon–ethane to helium–propane. With its hit resolution of 20 μm in the $r\phi$ projection and 25 μm in z the silicon had the potential of significantly improving the resolution for extrapolation of tracks into the vertex, and the new drift chamber gas with its 14% improved hit-on-track efficiency and reduced multiple scattering offered significantly improved momentum resolution. Understanding the new configuration and getting the ultimate efficiency and resolution was a considerable effort though. It led to a better understanding also of the earlier CLEO-2 tracking. This prompted a desire to reap the benefits of this improved understanding by repeating the event reconstruction for all the past CLEO-2 data and Monte Carlo simulations with the updated tracking software. So eventually we had three data sets to compare: the original CLEO-2, CLEO-2-recompress and CLEO-2.5. At first there were significant disagreements among all three in event efficiencies — and for a while the two newer data sets did not look so good. This had to be understood during a time when most CLEO members were involved in building CLEO-3. It took much longer than anyone anticipated, but was eventually accomplished. As a result, the publication of many CLEO-2.5 data analyses was delayed, and CLEO-2 data were still being studied many years beyond the start of CLEO-2.5 running.

The rapid growth of the collaboration slowed. It peaked in 1996 with 212 authors on the CLEO papers (see Table 4). Although new members joined in later years (Table 2), the outflow to other collaborations, mainly westward to BaBar and Belle, caused the membership to plateau for a while and then to decline slowly.

With so many members and so much work to do, the management of the collaboration became complex enough to require a change from single spokesman to two co-spokesmen, starting in 1997 with Ed Thorndike and George Brandenburg. Partitioning the leadership effort made it easier for non-Cornellians to take on the leadership responsibility. The CLEO data taking shifts evolved from two CLEO members for each of the three shifts, to one per daytime shift (plus two on each of the other two shifts), eventually to one CLEO physicist per shift plus one hired technician to handle the routine operations. In December 1999 we celebrated 20 years of CESR, CLEO and CHESS with invited outside speakers reciting the accomplishments in the Cornell Theater Arts Center, followed by an evening banquet in the Statler Ballroom. Twenty years of running is a very long time for any high energy physics collaboration, perhaps a record.

A nice demonstration of the vertex resolution of the silicon detector was Dave Cinabro's observation for the first time of the beam–beam pinch effect in the horizontal width of the beam at the collision point. The silicon data on displaced track vertices enabled CLEO to improve significantly on previous measurements of the D^+, D^0, D_s and τ lifetimes . CLEO also set a limit on $D^0 \leftrightarrow \overline{D^0}$ mixing, using the time dependence of the decays to separate the mixing from double Cabibbo-suppressed decays.

As before, however, the main thrust of the CLEO analysis effort was in the area of B meson decays, with data taken at the $\Upsilon(4S)$ resonance and just below $B\overline{B}$ threshold in 2:1 ratio. For most B analysis topics the new CLEO-2.5 data were combined with the data set available from the 5 fb of pre-silicon CLEO-2 integrated luminosity. Much of the published work (see Appendix Tables) in this period involved improvements in the accuracy of the measurements that fix the sides of the unitarity triangle in the complex plane representing the relation

$$V_{td}V_{tb}^* + V_{cd}V_{cb}^* + V_{ud}V_{ub}^* = 0$$

among the elements of the Cabibbo–Kobayashi–Maskawa matrix.

- $|V_{cb}|$ was obtained from the branching fraction for semileptonic B decays to charm — inclusively, and in the exclusive channels $B \to D\ell\nu$ and $D^*\ell\nu$.
- $|V_{ub}|$ came from the branching fraction for semileptonic B decays to non-charmed final states — inclusively from the tail of the lepton momentum spectrum beyond the end point for decays to charm and exclusively from $B \to \pi\ell\nu$ and $\rho\ell\nu$.
- $|V_{td}|$ could be obtained from measurements of the rate for $B^0 \leftrightarrow \overline{B^0}$ mixing. Here the LEP and Tevatron measurements of the time dependence of the oscillation and limits obtained for the $B_s \leftrightarrow \overline{B_s}$ oscillation eventually eclipsed the data from the CLEO time averaged measurements.
- $|V_{ts}|$ could be obtained from the rate for $B \to X_s\gamma$. This measurement was steadily improved with more data. Although it was viewed mainly as setting a limit on exotic high mass objects contributing in the loop, one could take the

Standard Model as given with W and t in the loop and get a determination of $|V_{ts}|$.

Each of these measurements reached a level such that the accuracy of the determination of the area of the unitarity triangle, upon which the strength of CP violation in K or B decays depends, was eventually limited by model uncertainty in the connections between experiment and $|V_{ij}|$.

Another major thrust of the CLEO analysis in the CLEO-2.5 period was measuring and setting limits for branching ratios and charge asymmetries of rare charmless hadronic B decays. CLEO discovered the decays to $K\pi$, $K\eta'$, $K\phi$, $K^*\eta$, $K^*\phi$, $\pi\pi$, $\pi\rho$ and $\pi\omega$, typically in several charge combinations. Measured branching fractions ranged from 8×10^{-5} for $K\eta'$ down to 4×10^{-6} for $\pi^+\pi^-$, and limits were obtained in many modes ranging as low as 2×10^{-6} for K^+K^-. These results set off a wave of theory papers discussing the decays in terms of amplitudes involving $b \rightarrow u$ tree diagrams, $b \rightarrow sg$ gluonic penguin loops, electromagnetic penguins, and occasionally W exchange or annihilation diagrams. Of special interest was the unexpectedly high $K\eta'$ rate, still not understood. The motive for much of this work was the possibility that at least some of these decays would show direct CP violation from the interference of tree and loop amplitudes. CLEO looked for asymmetries in five of these modes (and also in $\psi^{(\prime)}K^\pm$, $K^\pm\gamma$ and $X^\pm_{s,d}\gamma$) but did not see any at the 12 to 25% level of sensitivity.

In the charm decay sector CLEO-2 and -2.5 data provided precision measurements of key normalization modes in charm and bottom physics: $D \rightarrow K\pi$, $D \rightarrow K\pi\pi$, $D_s \rightarrow \phi\pi$ and the D^* to D decay modes. It was also possible to improve the measurement of the rate for $D_s \rightarrow \mu\nu$, an important check on lattice determination of decay constants. The collaboration continued its dominance of the field of charmed baryon spectroscopy with the discovery of over half of all known states of the Λ_c, Σ_c and Ξ_c.

With the world's largest sample of τ decays, CLEO specialized in rare and forbidden decay modes, with sensitivities in the $10^{-4} - 10^{-6}$ range. The large data sample also enabled the tau specialists in the collaboration to pursue a detailed exploration of hadronic spectral functions. However, efforts to improve the upper limit on the tau neutrino mass were disappointing.

16

Building CLEO-3, 1996–2000

———————————————— • ————————————————

The outstanding weakness of each CLEO detector has always been high momentum particle identification — in particular, distinguishing kaons from pions of the same momentum. There are three observables that can be used, at least in principle, to measure particle velocity: time of flight, ionization and the Cherenkov effect. Once velocity and momentum are known, mass follows from $m = p\sqrt{1 - \beta^2}/\beta c$. The problem for each of these techniques is that at high momentum, β gets immeasurably close to one, whatever the mass is. One therefore has to measure flight time ($= L/\beta c$), ionization ($\sim \text{const}/\beta^2$), and/or Cherenkov angle ($\cos\theta = 1/\beta n$) to very high precision, and over most of the solid angle. For CLEO-2.5 (or -2), the $K - \pi$ separation in ionization at the 2.5 GeV/c momenta important for distinguishing $B \to K\pi$ from $B \to \pi\pi$ was only 2.0 (or 1.7) standard deviations, and the resolution in time of flight at that momentum was useless. Three parallel R&D efforts were carried out to find a better solution: an aerogel threshold Cherenkov counter system, a high pressure sulfur-hexafluoride gas threshold Cherenkov counter array, and a ring imaging Cherenkov counter. The latter (RICH) appeared to be best able to provide at least three standard deviations of $K - \pi$ separation over the full momentum range. Giving up the existing time of flight counter array and reducing the outer radius of the drift chamber would provide enough radial space for a proximity focused RICH counter with photon detection by TEA, wires and cathode pads.

The other main motive for upgrading CLEO was the space interference between the existing drift chamber flat endplates and any significant improvement in the IR focusing quadrupoles. A new design with an endplate stepped inward for shorter wire length at smaller radii would allow us to install superconducting quadrupoles close to the interaction point (see Fig. 29). Tracking resolution could be maintained in spite of the reduced outer radius by replacing the inner silicon layers and the VD

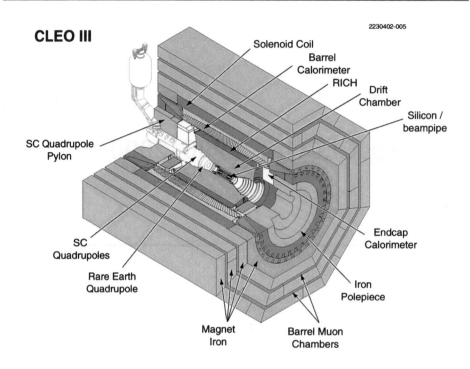

Fig. 29.   Cutaway view of the CLEO-3 detector.

chamber by a new, larger 4-layer silicon detector. This would be an opportunity to take advantage of recent advances in radiation hardening of silicon and replace the limited life 3-layer silicon detector before it died.

Cornell Senior Research Associate Chris Bebek had the job of managing the CLEO upgrade — budgeting, scheduling, coordinating parallel activities. The Syracuse group took the main responsibility for building the RICH detector with help from Southern Methodist, Albany and Wayne State. The detector was to be a cylindrical shell, comprising in radial order lithium fluoride crystal radiators, both planar and saw-toothed, a gas volume, calcium fluoride crystal windows, TEA + methane gas to produce photoionized charges, wires for multiplication, cathode pads and readout electronics. The main bottleneck turned out to be the production of LiF and CaF_2 plates. Optivac, the supplier, had problems with quality control and it was only with intensive intervention by the Syracuse crew (post-doc Ray Mountain, in particular) that enough radiator and window pieces were finally delivered within a year of the date originally foreseen. The RICH (Fig. 30), which had to go into CLEO first, was installed starting in June 1999. CLEO-2.5 data collection had already stopped in February 1999, the inner part of the detector had been dismantled, and the remainder had been running as a test bed for the new, faster readout and trigger electronics that had been developed by Ohio State, Illinois, Purdue and Caltech.

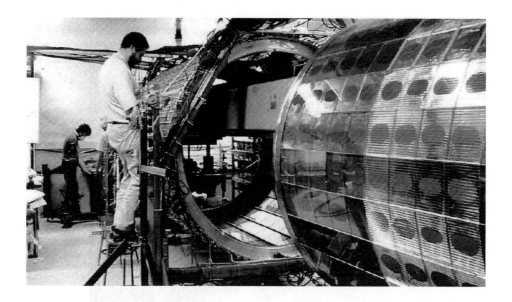

Fig. 30. Assembly of the Ring Imaging Cherenkov detector at Syracuse University.

The drift chamber went in immediately after the RICH. DR3, as it was called, was designed and assembled at Cornell under the direction of Cornell's Dan Peterson. Most of the wires were strung by hand; the innermost layers in the stepped "wedding cake" part of the endplate were strung by a robot constructed by the Vanderbilt group. Rochester provided the outer cathode z-strip layer.

The new silicon detector consisted of four layers of double-sided silicon wafers arranged cylindrically around the new 2.1 cm radius water-cooled, double-walled beryllium beam pipe. Ohio State, Cornell, Harvard, Kansas, Oklahoma and Purdue shared the job of producing the silicon detector and its electronics. It turned out to be a much more time consuming project than anyone anticipated. There were serious delays in component deliveries and assembly was labor intensive. The silicon detector was not ready for installation when the other major subsystems were installed. So CLEO-3 had an engineering run with a dummy in place of the silicon detector until the real silicon was ready for installation in February 2000. The shakedown run was not entirely wasted time, because DR3, the RICH, the data acquisition system, and the new C++ software all required a lot of tune-up and bug fixing. The new beam pipe that went in with the silicon was connected to the rest of the CESR vacuum with a cleverly designed remotely actuated "magic flange". Figure 29 shows a cutaway view of the new detector configuration. Figure 31 shows some of the CLEO collaborators posing by the completed detector.

Although the new CLEO-3 detector eventually worked well (see the event display in Fig. 32) and produced good physics results, it cannot be counted as a complete success. The delays hurt CLEO productivity at a time of intense competition with

Fig. 31. Some of the CLEO collaboration members and the newly completed CLEO-3 detector in 1997.

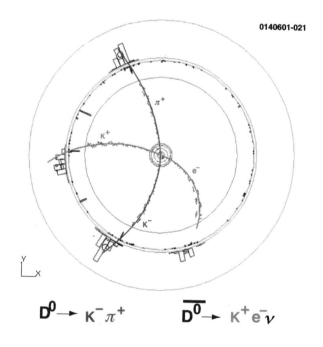

Fig. 32. Computer generated display of one of the early CLEO-3 events.

the BaBar and BELLE collaborations. Also, the silicon $r - \phi$ side efficiency began to degrade almost immediately, starting with the innermost layer and advancing outward. Something was definitely wrong with the batch of silicon we got from Hammamatsu; it seemed to be extraordinarily sensitive to radiation, although our dosimeters in the interaction region were telling us that the dosage was actually well within safe limits.

17

Phase III CESR Upgrade, 1996–2001

The Phase III CESR upgrade was proposed to the NSF immediately following the unsuccessful competition with SLAC for DOE funding for an asymmetric B Factory. Phase III, following on the earlier phase II, was the much less expensive "plan B" alternative. The goal was to improve CESR luminosity so that it could compete with the PEP-2 and KEK-B rings on the many physics topics that did not require asymmetric beam energies. At the same time we planned to increase the number of available synchrotron radiation user stations by instrumenting the back-fire radiation from the opposite-sign beam passing through the wiggler magnet that generated the X-ray beams for the A line. This new G-line was to be taken out to a separate new experimental area built into the hillside west of Wilson Lab.

The phase II interaction region focusing on the small-angle beam crossing configuration installed in 1995 and the complement of four 5-cell normal conducting rf cavities were optimal for colliding beams of 18 bunches per beam, that is, two bunches spaced by 42 ns in each of nine trains spaced by 284 ns. In this condition CESR had reached beam luminosities of $4.4 \times 10^{32}/\mathrm{cm}^2\mathrm{sec}$ with currents of 180 mA per beam and a beam–beam tune shift of $\xi_v = 0.041$. The stated goal of phase III was to increase the peak luminosity to at least 1.7×10^{33}. This was to be done by increasing the number of circulating bunches in each beam to 45, keeping the charge per bunch about the same. The separated pretzel orbits required to accommodate the bunches without parasitic collisions are diagrammed in Fig. 27.

In order to handle the higher beam currents we had to complete the following:

- replace the copper rf cavities with four single-cell superconducting cavities,
- replace the focusing quadrupoles nearest the collision point with stronger, superconducting magnets,

105

Fig. 33.   Cutaway view of the single-cell superconducting cavity in its cryostat.

- refurbish the linac injector to provide higher positron currents more reliably,
- upgrade the vacuum in the interaction region,
- upgrade the feedback systems for beam stabilization.

Most of the input power in a normal copper accelerating cavity is wasted in I^2R heating losses in the cavity walls. In order to minimize the power level for a given accelerating field, the beam aperture has to be as small as it is in the magnets. This unfortunately also facilitates the trapping of higher mode parasitic fields in the cavity by the passage of the short beam bunches. Higher order mode fields can destabilize the beam and limit the achievable beam current. In superconducting rf cavities the wall losses are negligible, so the beam aperture can be made much larger. Beam-cavity higher mode coupling is reduced and the highest trapped frequency is lowered.

To get some operational experience with the new system the SRF group installed in September, 1997, the first of four single-cell srf cavities (see Fig. 33) in place of one of the 5-cell nrf systems. It operated well with beam currents up to 360 mA. Three more niobium cavities were delivered in November 1997. Fabrication of the four cryostats took longer but was eventually accomplished. The assembly, testing and installation of the four cavities took place serially, the last one coming into operation in September 1999. Meanwhile we completed a major cryoplant — three big helium refrigerator–compress sets installed in a new room excavated under the transformer pad at the Kite Hill upper entrance to the Lab. We also acquired two new klystron power supplies designed by SLAC. Although there were some early problems with power limiting phenomena — window arcing, cavity surface defects,

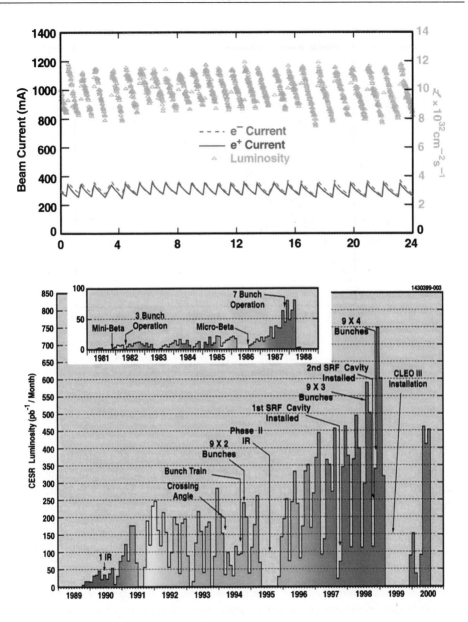

Fig. 34. (top) Scoreboard showing 24 hours of instantaneous luminosity, positron and electron beam currents, for a better-than-typical day. (bottom) Integrated CESR luminosity per month.

vacuum leaks — the srf system eventually turned out to be as reliable as the previous NRF. Beam instabilities were still encountered, but at higher beam currents than before. By early 2001 (before the superconducting IR quad installation) over 350 mA were circulating in 45 bunches in each beam and the peak luminosity had exceeded 1.2×10^{33} (see Figs. 34 and 35).

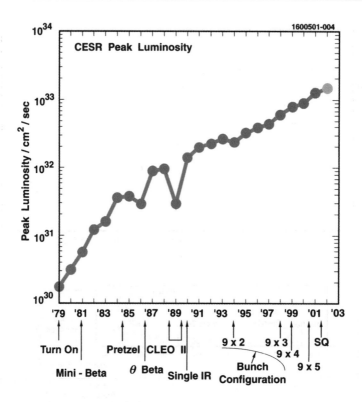

Fig. 35.   Evolution of the annual highest CESR peak luminosity.

When two bunches are colliding at the interaction point, the following e^+ bunch and the preceding e^- bunch pass each other at a transverse separation equal to the product of the longitudinal bunch spacing and the crossing angle. The bunches perturb each other and limit the achievable beam current. The effect is proportional to the magnitude of the β function at that point in the orbit. This occurs near the location of the interaction region quadrupoles where β can be quite large. To minimize the perturbation one needs to get as much focusing strength as possible as close to the interaction point as possible. In phase II the innermost focusing elements are permanent magnets. The phase III design replaces most of the permanent quadrupole with superconducting quadrupoles having much stronger gradient. This reduces the maximum of the β function, pushes it closer to the IP, and thereby decreases the beam–beam disturbance.

The new focus system had four combination quadrupole, skew-quadrupole, and steering dipole magnets in two cryostats. They were built by Tesla Ltd. from designs made in consultation with CERN and Cornell. Personnel turnover and inexperience at Tesla delayed delivery and testing of the completed magnets and cryostats beyond the day when the last of the CLEO-3 components were installed. To give CLEO-3 enough integrated luminosity to make a good showing at the summer 2001 conferences we decided to postpone the phase III quadrupole installation shutdown until June 2001. In the same summer 2001 shutdown we finished the installation of the special magnet components for the new G-line for synchrotron radiation users. We also put in a new positron production target in the linac, with better alignment control and stronger solenoid focusing. This immediately halved the positron injection time.

18

The Money

•

In the CESR–CLEO era most of the financial support (typically 97%) of the Cornell Laboratory of Nuclear Studies came from the Elementary Particle Physics program of the Physics Division of the National Science Foundation, and over 90% of that was in the form of a Cooperative Agreement for the support of CESR and CLEO. The remainder consisted of grants to support the Lab theory group and occasionally other activities. The Cooperative Agreement was something between a grant and a contract, specially devised for the support of CESR and CLEO.

The Cornell Electron–positron Storage Ring is the only high energy accelerator facility supported by the NSF. The other accelerators in the US are funded by the Department of Energy, as are most of the university user groups. Fortunately for us, our keepers in the NSF Physics Division have always felt a strong obligation to keep CESR and CLEO healthy as long as they remained productive. Thanks to the enthusiastic reports from the consultants that the NSF picked to review our proposals and the goodwill that our support of the CHESS X-ray science program generated in other divisions of the NSF, the Physics Division succeeded in getting enough in the annual NSF apportionment of the Congressional appropriation to make it possible to keep CESR and CLEO alive. I say alive, because we never got the entire amount of our annual proposal.

Most of the CLEO groups from other universities were supported by the DOE. A minority were funded by NSF, and the Canadians got their support from NSERC. These separate grants to the collaborating groups covered their salaries, travel, at-home computing expenses, and sometimes minor equipment contributed to the CLEO detector. The Cornell CESR–CLEO account normally paid for major equipment constructed by collaborators, for example, during detector upgrades.

The CESR–CLEO annual base budget increased steadily from $5 million in 1980 to $20 million in 1999, keeping ahead of inflation and allowing us to maintain operation of a facility that was steadily increasing in power and complexity. This base support was supplemented for a few years in each decade by a pulse of extra funding for new equipment: $20 million in 1977–1979 for the original CESR and CLEO-1 construction, $37 million in 1985–1989 for the CESR and CLEO-2 upgrade, and $27 million in 1994–1998 for the CESR and CLEO-3 upgrade. In a typical non-upgrade year the base budget went 46% for salaries and fringe benefits, 15% for facility operations — mainly CESR supplies and power, 5% for CLEO research supplies, 11% for capital equipment and 23% for indirect costs to Cornell. Indirect costs are a percentage tax on all university research direct costs that pays for everything from libraries to lawn mowing.

The number of employees supported on the CESR–CLEO Cooperative Agreement remained rather stable over the decades, increasing from a total of about 144 in 1980 to 157 in 1997. Throughout the period the breakdown was always close to 13 faculty (paid by the Lab in the summer and by the Cornell Physics Department during the academic year), 29 PhD Research Associates, 30 engineers, 53 technicians, 21 grad students, and 11 service personnel. In addition there were part-time undergraduate assistants and short-term help hired during upgrades. The steadiness of the annual base funding meant that it was never necessary to lay off a regular Laboratory employee for lack of NSF support. On the other hand, hiring a new employee was hardly ever a viable way to get a new task accomplished.

Although the annual funding from the NSF was never drastically cut, there were periods of anxiety. Since the CESR–CLEO Cooperative Agreement represented about 40% of the budget of the NSF Elementary Particle Physics program, it was impossible to insulate us from the year by year fluctuations in the Congressional NSF appropriation and in the allocation that trickled down to the Physics Division. Even with an approved annual budget proposal we could never be sure how much we would receive until the fiscal year actually started, or sometimes even later. This meant frequent submission of itemized budget amendments to respond to changes in the bottom line.

It was not as bad as it may appear, however. The Laboratory of Nuclear Studies was small enough so that it was possible to exercise tight, centralized control of spending. In hard times we could postpone or slow down purchases of supplies and equipment. Since the NSF Physics Division had no more than three program officers for all of high energy physics, it was impossible for them to micromanage the CESR–CLEO program, so we had a reasonable degree of flexibility in programming our finances. The NSF allowed year-to-year carryover of unspent funds, so we put aside a few percent of the annual budget, spending it only in an emergency. The NSF also allowed any defensible shifting of funds among categories when our plans had to deviate from what was laid out in our five-year proposal. By government rules we had to consider the money spent as soon as it was committed, for example, when we placed an order. However the Cornell Sponsored Programs Office, which held and disbursed the dollars, knew the money was spent only when the bill was actually paid. In effect, this allowed us to "borrow" a few months ahead, in the case of a temporary shortfall. On average, we lived within our means, and we were able to handle fluctuations without undue hardship.

19

A New Director and a New Direction, 2000–...

———————————— • ————————————

My third five-year term as Director of the Cornell Laboratory of Nuclear Studies was due to expire on 30 June 2000. As the SLAC and KEK B factory projects started to produce physics results, I saw that the Laboratory would soon have to make a major shift in direction in order to stay viable. I could see my retirement from research coming in a few years, and it seemed to me that someone more likely to be an active participant in the future course of the Lab should be the dominant voice in deciding what the course should be. So I announced in summer 1999 that I would not be taking another term as director.

The Cornell Vice President for Research appointed Prof. Persis Drell to head a search committee for a new director. The committee solicited suggestions of candidates from inside or outside the Lab. Maury Tigner was on everyone's list, even though he had retired back in 1995 for reasons of health. He was certainly best qualified for the job: he pioneered the rf cavity design used in all high energy electron synchrotrons, he initiated the development of superconducting rf cavities, he was responsible for the conceptual design of CESR and had been the project manager for its construction, he had run the Central Design Group for the SSC. He has the same qualities — accelerator expertise, creativity, competitive spirit, and management know-how — that enabled Bob Wilson to launch the Cornell Laboratory on its successful career at the frontier of elementary particle physics. Back in 1995 Maury had a serious operation on his spine, though, and the doctors had apparently discouraged him from going back to work. On the off chance that Maury might have second thoughts about retirement, the committee asked if he would consider the directorship. It was not such a far-fetched idea, since he had been spending much of the intervening time working at the Beijing laboratory, helping them with plans to upgrade their machine. He accepted. Although I had originally had the idea of a younger replacement, this was an ideal outcome. I could resign with a clear conscience, knowing that the Lab would be in good hands and would have the best possible chance of weathering the coming storm.

The storm was the competition from the new asymmetric B Factories coming into being. CESR was a victim of its own success. The exciting field of flavor physics opened up by the CLEO discoveries of the last decade, and the demonstration at CESR that one could reach luminosities well above 10^{32} cm^{-2}s^{-1} had inspired the other laboratories to follow our lead. They were much larger and had more resources, however. It was clear by 2000 that they were going to meet their luminosity goals, in spite of the complexities inherent in asymmetric energies. CESR would have a hard time keeping pace. For the past year or two we had been actively considering a Phase IV CESR — an attempt to leapfrog PEP-2 and KEK-B in luminosity to reach 10^{34}. Phase IV would have involved building a new dual-aperture equal-energy storage ring on top of the synchrotron, turning the old CESR over to CHESS. Alexander Mikhailichenko developed and prototyped a new dual-aperture superconducting quadrupole, and Joe Rogers laid out a design for new bending magnets and vacuum chamber. In early 2000, though, we began to consider seriously other options for maintaining physics productivity in the next decade.

- Join an existing collaboration at another laboratory — say LHC, BTeV or BaBar. This was not a popular idea. It was already too late to have much of an impact on the important decisions. Also, there would be no role for the accelerator physicists at the Lab.
- Run above the $\Upsilon(4S)$ concentrating on $B_s - \overline{B_s}$ production at the $\Upsilon(5S)$, for instance. This idea did not catch on either. The $\Upsilon(5S)$ does not stand out much above non-$b\bar{b}$ background, and the $\Upsilon(5S)$ peak is mainly B states, not B_s. The competition from experiments at hadron machines with their much higher production rates would likely overwhelm us.
- Run on the bound state resonances $\Upsilon(1S)$, $\Upsilon(2S)$, $\Upsilon(3S)$. This topic has had very little attention since the early 1980s, soon after the upsilons were discovered. The strong interaction is responsible for their binding and decay, so they tell us about QCD. No one expects to see surprises or upset QCD, so ARGUS, CLEO, BaBar and BELLE have concentrated instead on B physics at the $\Upsilon(4S)$. However, there are theorists anxious to test nonperturbative strong interaction calculation techniques, most notably lattice QCD. New experimental results would challenge them to improve their approximations. Where are the intermediate D states (Fig. 36)? Where are the singlet states η_b, η_b' and h_b? Can we understand the pattern of hadronic decay modes? Most of us felt that it would be worth our while to spend about a year running on the resonances. There did not seem to be a viable long term program here, though.

- Run CESR in the tau–charm threshold region. There has been a history of unsuccessful proposals for tau–charm facilities elsewhere. There were reasons for the lack of enthusiasm. Charm physics has been considered less interesting than b-quark physics, because of the structure of the CKM matrix. Charm decays involve only the upper left 2×2 corner of the matrix, which depends on just the Cabibbo angle and has no lowest-order imaginary component to give rise to a measurable Standard Model CP violation. The c-quark can decay to its weak-isospin partner, the s-quark, while the corresponding $b \rightarrow t$ transition is energetically forbidden. So although rare b-decay processes can compete with the suppressed $b \rightarrow c$ transition, the corresponding rare c-decay processes cannot compete so well with the nonsuppressed $c \rightarrow s$ transition. Moreover, most of charm physics — and also tau physics — is available for free when you run on the $\Upsilon(4S)$. Since the first tau–charm factories were proposed, many of the early physics motivations have since been accomplished by CLEO and other experiments at or above the b-quark threshold.

In spite of the gloomy history of tau–charm proposals, the latter option was the one we decided on. Our Syracuse University collaborators, Sheldon Stone and Marina Artuso, convinced the skeptics that the situation had changed. The interpretation of the B decay data in terms of basic weak interactions of the b-quark depends critically on the understanding of the strong interaction effects — binding, rescattering, gluon processes, form factors, strong phases, and such. Accurate measurements of D decay processes, where the strong effects were larger and the weak interaction physics was supposedly well understood, would allow theorists to test and refine their nonperturbative approximation techniques and thereby put b-physics on a reliable quantitative footing. The most important advantage of doing charm physics near $D\bar{D}$ threshold rather than at $B\bar{B}$ threshold would be the cleanliness of the $D\bar{D}$ final states at threshold. One could expect to tag at least 20% of the decays and thus make a substantial impact on lowering systematic errors in branching ratio measurements.

In order to run CESR at the charm threshold — one third the usual CESR energy — and have enough luminosity to produce $D\bar{D}$ at a rate comparable to the $D\bar{D}$ rate at b-threshold, several requirements would have to be met.

- The CESR magnet guide field, including the focusing in the interaction region, would have to scale with energy. Thus the major part of the permanent magnet final focus quadrupoles would have to be replaced by an electromagnet. The installation of the phase III upgrade superconducting quadrupoles would accomplish this. This would at the same time enable us to gain luminosity by reducing the β_y^* at the interaction point. Since the luminosity ($\propto E^2/\beta_y^*$) tends to decrease at lower beam energies, one will have to regain as much of that as possible.

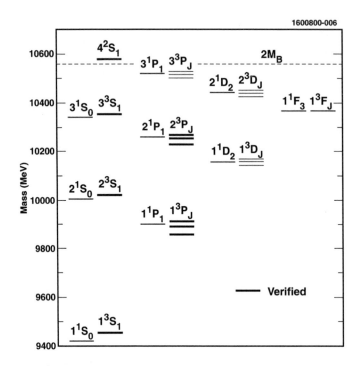

Fig. 36.   Energy levels in the upsilon bound state system.

- Because of the hour-glass effect, the luminosity gain with low β_y^* will come only if the bunch length is shortened to match β_y^*. That requires high rf cavity voltage. The superconducting cavities, including the two additional ones on order, were actually ideal for this. Instead of providing mainly for the power radiated at high beam energies, they would be shortening the bunches at low beam energies.
- At the lower energies the radiation damping of the transverse oscillations of the beam particles is rather ineffective. One has to introduce wiggler magnets to shorten the characteristic damping time and thus keep the transverse size of the beam small for high luminosity. Fourteen 1.33-m long superferric wigglers spread around the CESR lattice would do the trick. In terms of money and effort this would be the only major hurdle in turning CESR into CESR-c; that is, about $4 million and 2 years. In the meantime we could be running on the $\Upsilon(1S)$, $\Upsilon(2S)$ and $\Upsilon(3S)$. Figure 36 shows the current status of the energy levels in upsilon spectroscopy.

20

Concluding Remarks

---•---

The Cornell Electron Storage Ring and the CLEO experiment have been quite successful. For a while CESR boasted the world's highest colliding beam luminosity, the first to exceed $\mathcal{L}_{pk} = 3 \times 10^{32}/\text{cm}^2\text{s}$ (see Fig. 35). The $\Upsilon(4S)$ resonance has been the ideal energy for studying the properties and interactions of the B_d and B_u mesons, and for 21 years the CLEO collaboration was the leader in heavy quark and lepton physics. The $\Upsilon(4S)$, $\Upsilon(5S)$, $\Upsilon(6S)$, $\chi_b(1P)$, $\chi_b(2P)$, B, B^* and D_s mesons were discovered at CESR, as were the transitions $b \to c$, $b \to u$ and $b \to s$. As of 2001 over half of the entries in the Particle Data Group tables for B mesons and for charmed mesons and baryons were based primarily on CESR results.

Largely through the research effort at CESR, we have learned the following facts about heavy quarks and leptons in the Standard Model.

1. The spectroscopy of bound $b\bar{b}$ states confirms the expectations of quantum chromodynamics. The perturbative and nonperturbative predictions are nicely confirmed by the masses of the Υ and χ_b states, the radiative transition rates between the states, and the value of the strong coupling α_S derived from the $\Upsilon \to gg\gamma$ and $\Upsilon \to ggg$ decay rates.

2. The decays of the B mesons support the Kobayashi–Maskawa picture of six-quark universality in the charged-current weak interaction. Heavy Quark Effective Theory and models based on factorization provide an adequate description of a wide range of data. Before the t quark was discovered, the b quark data pointed to its existence and gave indications of its mass. Our information on the values of the CKM matrix elements $|V_{cb}|$, $|V_{ub}|$, $|V_{ts}|$ and $|V_{td}|$ comes from data on B decays and $B\bar{B}$ mixing.

3. The discovery of the $b \to u$ transition established that all of the CKM matrix elements are nonzero, thus allowing CP violation in B decay. This determination of $|V_{ub}|$ and the measurements of $|V_{cb}|$ and the $B^0\overline{B^0}$ mixing rate give us the three sides of the unitarity triangle, and challenge us to measure the three angles through the observation of CP asymmetries. Agreement between the sides and angles should tell us whether the Kobayashi–Maskawa mechanism (a complex element in the CKM matrix) is sufficient to account for CP violation.

4. We have now available an extensive data base on the strong, electromagnetic, and weak interactions of particles containing the c and b quarks, that can be used to test speculations on new physics beyond the Standard Model. So far, though, there is no statistically significant evidence for new physics.

5. The decays of the τ lepton confirm e, μ, τ universality.

In spite of the successes, there are still puzzles. There is no compelling explanation for the double-humped spectrum of $\pi\pi$ invariant masses in the $\Upsilon(3S) \to \Upsilon(1S)\pi\pi$ transition, nor for the unexpectedly large branching ratio for charmless $B \to \eta' X$ (and $\eta' K$). There is no understanding of the pattern of quark and lepton masses and CKM matrix elements, nor an answer to the question of why there are six quarks and six leptons. And we do not know whether the baryon–antibaryon asymmetry in the universe requires another source of CP violation.

What has been CESR's secret of success? How has the Cornell laboratory managed to prosper when practically all of the once numerous university facilities in high energy physics have closed down — Caltech, Carnegie, Chicago, Columbia, Harvard-MIT, Princeton-Penn, Rochester, Purdue? There are several reasons, I believe.

People. Bob Wilson was a great physicist, a clever inventor, an inspiring leader, and he had great ambitions for the Laboratory. With this rare combination of gifts he launched the Laboratory on the course it has followed ever since. His can-do attitude and commitment to keeping the Lab at the forefront infected everyone, even for decades after he left. And it attracted other creative minds to the enterprise — McDaniel, Littauer, Tigner, Siemann — just to name a few. CESR would not have been possible without the leadership of McDaniel and the inspiration and project direction of Maury Tigner. But leadership is not everything. Many times in the past, outsiders have expected the Lab to falter when one of the big names left the scene. Each time, the Lab demonstrated a depth and breadth of talent, experience and commitment sufficient to carry on in the established Cornell tradition. In CLEO, which is dependent on the contributions of many people from many institutions working together, the democratic structure of the collaboration has been an important factor. Everyone works hard and enthusiastically because everyone participates in the decisions.

Innovation. A crucial aspect of the Cornell tradition is the continual renewal of the accelerator and the experiments, keeping them productive at the physics frontier. For fifty years the Lab has built a new accelerator — 300 MeV, 1 GeV, 2 GeV, 10 GeV, CESR — or started a major upgrade — as in 1985 and 1994 — every 8 or 9 years. It has always been considered important to look far enough ahead to prepare for the time when the current capabilities are no longer exciting enough to justify support.

Focus. For decades the top priority of the Lab has been the performance of CESR and its experimental program. There are no other priorities. During the 1980s CESR competed on a par with DESY, a lab with an order of magnitude advantage in resources. CLEO eventually outperformed ARGUS so decisively that DESY gave up the competition. The main reason was that DESY was never able to devote its full attention to the DORIS ring and the ARGUS experiment. DORIS had to play second fiddle to the HERA project.

Cost Conciousness. One of Wilson's legacies is the impulse to save money by being clever. As a result, Cornell and CESR have always had a reputation for delivering the most for the least. Of all the 1991–1999 *Phys. Rev. D* and *Phys. Rev. Lett.* papers based on HEP experiments at BNL, CESR, Fermilab, and SLAC, 23% have come from CESR, while CESR has accounted for only 3.9% of the total HEP spending of the four labs. The ability to upgrade the facility periodically has depended on matching the Lab appetite to the capabilities of the NSF to provide funding. It is a fact that no capital project proposal requiring less than a doubling of the annual funding level has been refused, and also that no proposal that required more than a doubling has been accepted. So we can credit the NSF as well as Wilson for the Lab's parsimony.

The NSF. There is no escaping the fact that the trust and generosity of the Physics Division of the NSF have been crucial to the success of CESR and CLEO. We owe an enormous debt to the Division directors and program officers over the years: Al Abashian, Marcel Bardon, David Berley, Chuck Brown, Willi Chinowsky, Joe Dehmer, Bob Eisenstein, Alex Firestone, Norman Gelfand and Tricia Rankin among others. They considered CESR the flagship of the NSF program in high energy physics, and worked hard to keep it afloat.

Luck. While we are thanking people, we have to remember that Mother Nature and Lady Luck have been extraordinarily kind to CESR. Although the CESR energy was fixed by the existing tunnel length before the discovery of the b quark, the energy turned out to be just right for covering the threshold region for $b\bar{b}$ production, from the $\Upsilon(1S)$ to the $\Upsilon(6S)$ and beyond. Other e^+e^- machines built around the same time (DORIS, PETRA, PEP, TRISTAN) have all ceased to produce useful physics, because their energy choices were not as lucky. The fortuitous occurrence of the $\Upsilon(4S)$ resonance just above $B\bar{B}$ threshold was ideal for producing B^+B^- and $B^0\overline{B^0}$ copiously and cleanly. The value of V_{cb} was small enough for rare processes, such as $b \to u$, $b \to s$, and $b\bar{d} \leftrightarrow \bar{b}d$, to compete with the dominant $b \to c$ decay. And V_{ub} is nonzero, allowing for the possiblility of CP violation in B decay.

The primary goals for the CESR facility and the CLEO experiment are clear.

1. Continue important, productive research in heavy quark and lepton physics as long as it is interesting as a window on the Standard Model and beyond: CP violation, rare loop decays, leptonic decays, tagged studies, heavy meson and baryon spectroscopy, charm decays, rare τ decays, and so on.

2. Continue to improve and extend CESR performance, not only to advance the CESR/CLEO HEP goals but also to serve the worldwide accelerator community as a testbed for innovations.

3. Continue to serve the US high energy physics program by providing a user-friendly facility for faculty, post-docs, and graduate students from many universities to pursue world-class research in particle physics.

4. Continue to provide high intensity X-ray beams for the hundreds of users of the Cornell High Energy Synchrotron Source (CHESS).

21

Glossary

•

Accelerator: A machine that uses electric and magnetic forces to increase the energy of charged particles. Circular accelerators include cyclotron, synchrocyclotron, betatron and synchrotron; there are also linear accelerators.

AdA and Adone: Electron–positron storage rings at the Frascati National Laboratory, Italy. AdA, completed in 1961, operated at 0.2 GeV per beam, and Adone, completed in 1969, operated at a maximum energy of 1.55 GeV per beam.

Baryon: A hadron that is either a proton or a heavier hadron that ultimately decays into a final state that includes a proton. Examples include the neutron and several varieties of hyperon. A baryon is a bound state of three quarks.

BEPC and BES: The 2.8 GeV per beam Beijing Electron Positron Collider and its major detector. BEPC and BES began operation in 1988.

Betatron: A type of circular electron accelerator that depends on the rate of increase of magnetic field to provide the accelerating force.

Bremsstrahlung: The process of photon production by the deceleration or deflection of electrons in some target material. Photons used in high energy collisions are usually generated through Bremsstrahlung, starting with an electron beam from a linac or synchrotron.

Branching ratio: The fraction of the decays of an unstable particle that result in the specified final state.

CERN: The European Center for Nuclear Research in Geneva, Switzerland. Since its founding in the 1950's CERN has constructed and operated many accelerators, among them the SynchroCyclotron, the 28 GeV Proton Synchrotron, the 31 GeV per beam Intersecting Storage Ring, the 450 GeV

Super Proton Synchrotron, the 450 GeV per beam SP$\overline{\text{P}}$S proton–antiproton storage ring, and the 104 GeV per beam LEP electron–positron storage ring. CERN is constructing a 7 TeV per beam proton–proton storage ring, the LHC.

CESR: The 6 GeV per beam electron–positron storage ring at Cornell University, operating since 1979.

Charm: In order of increasing mass, the fourth of the six quark species. The charge is $+2/3 \times e$ and the mass is around 1.9 GeV. Charm also refers to the "flavor" or quantum number associated with the charm quark.

CKM matrix: The 3×3 matrix of weak decay couplings V_{ij} that link the various flavors of charge 2/3 quarks i and charge $-1/3$ quarks j. The CKM stands for Cabibbo, who first proposed the early 2×2 form of the matrix, and for Kobayashi and Maskawa, who suggested that the phenomenon of CP violation could be explained if there existed a third pair of quarks. The complex values of the matrix elements V_{ij} determine decay rates, mixing rates and the violation of CP in weak decays of mesons and baryons.

CLEO: The major detector facility at the Cornell Electron–positron Storage Ring and the collaboration that operates the detector.

Collider: An accelerator facility in which oppositely directed beams collide with each other. A collider can be either circular or linear. A circular collider for particle and antiparticle (e.g. CESR) can be a single ring in which the two beams travel on the same (or almost the same) orbit in opposite directions. Or there can be two intersecting rings, with one beam in each ring.

CP: is the operation that combines particle–antiparticle flip C and space inversion P. The universe is almost symmetric under CP, that is, the corresponding quantum number is conserved in strong and electromagnetic interactions. The weak interaction, however, can violate CP symmetry. That is, a particle decay rate, say for $B^+ \to K^+\pi^0$, can differ from the antiparticle decay rate, for $B^- \to K^-\pi^0$.

Cross-section: A measure of the probability of a specified particle + particle collision process. The cross-section has units of area and is dependent on the basic physics of the specified process. Multiplying the cross section by the accelerator-dependent luminosity gives the rate of interactions.

Cyclotron: A particle accelerator in which the particles (usually protons) circulate at gradually increasing radii in a fixed magnetic field. Each time they cross a gap between electrodes (called dees) they get a push from the electric field.

Decay: The transmutation of an unstable particle into several particles of lower total mass. This can happen through the strong, electromagnetic, or weak interaction.

Detector: A device that records the passage of a high energy particle. It can mean either an individual device, such as a scintillator, a drift chamber, a proportional tube, a silicon chip; or it can mean a multipurpose array of many such individual devices. Modern collider experiments use such detectors in the latter sense. The typical collider detector includes a drift chamber in a magnetic field to record the trajectories of all (or most) of the charged particles emerging from the beam–beam collision, along with other detector devices spread over the available solid angle to detect neutrals and identify as many of the produced particles as possible.

DESY: The German Synchrotron Laboratory in Hamburg. Since its founding in 1959 DESY has built and operated several accelerator facilities, among them the 7.5 GeV DESY electron synchrotron, the 5.3 GeV per beam DORIS electron–positron storage ring, the 22 GeV per beam PETRA electron–positron storage ring and the HERA 30×820 GeV electron–proton storage ring.

DOE: The US Department of Energy, the major funding source for US accelerator laboratories. It was preceded by Energy Research and Development Agency (ERDA), and before that the Atomic Energy Commission (AEC).

DORIS: A 5.3 GeV per beam electron–positron storage ring at DESY. **DASP** was the detector built and operated by the DASP collaboration from 1974 to 1978, then taken over by the collaboration that from 1982 to 1992 operated the **ARGUS** detector. Other detectors at DORIS were **PLUTO, DESY-Heidelburg, LENA** and **Crystal Ball**.

Drift chamber: A particle detection device that records the trajectories of charged particles. It operates by collecting on oppositely charged wires the ions produced in a gas by the passage of the particles. The measurement of ion drift times for a sequence of wires allows one to infer the distance between each of the wires and the particle track and thus reconstruct accurately the particle trajectory. If the drift chamber is in a magnetic field, the curvature of the particle trajectory is a measure of its momentum.

Electroproduction: The inelastic scattering of an electon off a nucleon in which one or several mesons are produced.

Fermilab: The national laboratory for high energy physics in Batavia, IL. It includes a 1 TeV proton synchrotron and a 1 TeV per beam proton–antiproton collider called the Tevatron.

Gluon: The virtual particle that carries the strong force. Quarks are bound into a hadron, such as a proton or meson, by the exchange of gluons.

Hadron: A strongly interacting bound state of quarks and/or antiquarks. Examples include the proton, neutron, mesons, hyperons.

Hyperon: An unstable baryon containing one or several heavy quarks: strange, charm or b. Strange hyperons include the $\Lambda = uds$, $\Sigma = uus$, etc., $\Xi = uss$, etc. and $\Omega = sss$.

KEK: The Japanese national high energy physics laboratory in Tsukuba. KEK has operated several accelerator facilities, including the 12 GeV proton synchrotron, the 32 GeV per beam TRISTAN electron–positron storage ring, and the 3.5 × 8 GeV KEK-B asymmetric electron–positron storage ring.

LEP: The 104 GeV per beam Large Electron–Positron storage ring operating at CERN from 1989 to 2000. The four detectors were **ALEPH, DELPHI, L3** and **OPAL**.

Lepton: Any one of the following: electron, muon, tau or their corresponding neutrinos. Leptons are distinguished from hadrons in that they do not interact with the strong force and they do not contain quarks.

Linac: A linear accelerator in which charged particles gain energy in a single pass through a row of microwave cavities. A linac is often used to inject a low energy beam into a synchrotron or storage ring. The highest energy linac is the SLAC 2-mile linac with a maximum energy of about 50 GeV.

Luminosity: A measure of beam intensity. When multiplied by the reaction cross-section in cm^2, it gives the rate of reactions per second. The units of luminosity are therefore cm^{-2}sec^{-1}.

Meson: An unstable hadron made of a bound quark and antiquark. Examples are π, η, ρ, ω, η', ϕ, ψ, χ, Υ, ...

Mixing: The process in which a neutral meson oscillates to its antiparticle, for example, $K^0 \rightleftharpoons \bar{K}^0$ and $B^0 \rightleftharpoons \bar{B}^0$.

NSF: The US National Science Foundation, one of the funding agencies for basic science in the US.

PEP: A 15 GeV per beam electron–positron collider operating at the Stanford Linear Accelerator Center from 1980 to 1990. The major detectors were **Mark-II, HRS, DELCO** and **TPC**.

PEP-2: The rebuild of the PEP collider into a high luminosity asymmetric 3.1 × 9 GeV per beam electron–positron collider. The resident detector is called **BaBar**.

PETRA: A 22 GeV per beam electron–positron collider operating at DESY from 1978 to 1986. The major detector collaborations were **CELLO, JADE, Mark-J** and **TASSO**.

Photoproduction: The production of mesons in the collision of a high energy photon (quantum of electromagnetic radiation) with a target.

Pretzel Orbit: A scheme for circulating two counter-rotating beams in the same ring, keeping them from colliding except at a few chosen interaction points. Typically the two beam orbits weave about each other, and the particle bunches are synchronized so that they arrive simultaneously only at the desired crossing points.

QED and QCD: Quantum electrodynamics and quantum chromodynamics, the modern theories of electromagnetic interactions and of strong (nuclear or hadronic) interactions. QED and the theory of weak interactions have been unified as Electroweak Theory, which together with QCD form the Standard Model of particle interactions.

Quadrupole: The four-pole magnet configuration used for focusing particle beams. A single quadrupole will cause a beam to converge in one dimension (say horizontal) and diverge in the orthogonal dimension (say vertical). A pair of opposite quadrupoles can focus a beam in both dimensions. Modern accelerators use an alternating sequence of quadrupoles to keep the beam confined transversely.

Quark: One of six types (or "flavors") of strongly interacting fundamental particles called up, down, strange, charm, bottom, top, or u, d, s, c, b, t. The masses range from a few MeV up to about 170 GeV. They exist only in bound states: three quarks forming a baryon, quark and antiquark forming a meson.

Resonance: In particle physics, a temporary bound state of several particles. In time it decays to its constituents; in mass (rest energy) it has a range of values characterized by a central mass and a width that is inversely related to its mean lifetime. In the reaction cross-section for the collision of the constituents plotted against energy, a resonance appears as a peak centered at the mass value and having the characteristic width.

RF: Radio frequency. In a circular or linear accelerator the RF system is responsible for accelerating the particles. It consists of microwave cavities, through which the beam passes, fed by a microwave power source. In Cornell jargon SRF stands for superconducting RF, using niobium cavities cooled by liquid helium. NRF means normal-conducting copper cavities.

Scintillator: A particle detection device consisting of a clear liquid or plastic that produces a pulse of visible light from the deexcitation of struck atoms left in the track of a high energy charged particle passing through. The light pulse is generally detected by a photomultiplier tube. The location, timing and amplitude of the pulse can be used to identify the particle. Scintillators are often used in large arrays covering a major part of the solid angle surrounding the beam collision point.

Silicon detector: A particle detection device based on silicon diodes formed either in strips or pixels on a thin wafer. The passage of a charged particle through one of the diodes will cause an electrical pulse which can be used to locate accurately the particle in space and time. The amplitude of the pulse carries information about the type of particle. Silicon detectors are typically located very near the point of particle production, and can be used to reconstruct decay vertices with accuracies of tens of microns.

SLAC: The Stanford Linear Accelerator Center. The accelerator facilities at SLAC include the 50 GeV electron linac and the 50 GeV per beam electron–positron SLAC Linear Collider, as well as the SPEAR, PEP and PEP-2 electron–positron circular colliders.

SPEAR: The 4.1 GeV storage ring at SLAC, operated as an electron–positron collider from 1972 to 1988. Experiments included **Mark-I**, **Mark-II** and **Mark-III**, and the **Crystal Ball**.

Storage Ring: An accelerator ring in which a beam or two beams can circulate at fixed energy. Storage rings have two uses, either to provide colliding beams for high energy physics experiments, or to provide radiated X-ray beams for experiments in materials science, microbiology, medicine and other fields.

Strangeness: One of the six quark "flavors", conserved in strong and electromagnetic interactions, but not in weak interactions. K mesons and Λ and Σ hyperons are strange particles, i.e. contain an s or $\bar{s}$ quark and thus have nonzero strangeness.

Synchrocyclotron: A type of circular accelerator mainly for high energy protons. It works on the same principle as a cyclotron, except the frequency of the RF accelerating voltage varies through the acceleration cycle. Otherwise, the protons would get out of phase with the accelerating voltage as their relativistic velocities did not increase in proportion to their momenta.

Synchrotron: A type of circular accelerator for relativistic particles. The particles travel on a fixed radius circular orbit through a ring of magnets. The magnetic guide field cycles from low to high values in proportion to the increasing particle momentum.

Tevatron: The 1 TeV per beam proton–antiproton collider at Fermilab. The two major experimental detectors are called **CDF** and **DØ**.

Unitarity triangle: A diagram in the complex plane illustrating the relation of the CKM matrix elements implied by the unitarity condition $V_{td}V_{tb}^* + V_{cd}V_{cb}^* + V_{ud}V_{ub}^* = 0$.

Weak interaction: The fundamental interaction responsible for most radioactive decays. The interaction involves a virtual $W^\pm$ or Z^0 boson as an intermediate or exchanged particle emitted or absorbed by a quark or lepton. Neutrinos can interact only through the weak interaction.

22

Appendix

Table 1. Luminosity and major upgrades.

Year	$\mathcal{L}_{pk}/10^{30}$ cm^{-2}s^{-1}	$\int \mathcal{L}dt$ pb^{-1}	CESR upgrades	CLEO upgrades
1979	2	1	CESR completed	CLEO-1 partially complete
1980	3	8		
1981	8	17	2nd rf cavity, minibeta	completed DX, MU; sc coil
1982	12	90	muffin-tin srf test	
1983	16	60	separators, 3 bunches	started DR2 construction
1984	37	104		VD, new DR electronics
1985	39	143	e^+ topping	
1986	30	96	microbeta REC quads	installed DR2, IV
1987	92	420	7 bunches	
1988	100	160	higher linac energy	μVD test
1989	30	45		CLEO-2 detector
1990	150	394	single IR	
1991	220	1100		
1992	250	1470		
1993	290	1390	5-cell rf cavities	
1994	250	1370	2 mr crossing, 18 bunches	
1995	320	816	new IR focusing	3-layer Si detector, He in DR
1996	400	2690	e^+ target	
1997	470	3400	SRF cavity in E2	
1998	720	4442	SRF in E1, 36 bunches	
1999	820	1010	SRF in W1,2	CLEO-3 DR3, RICH, DAQ
2000	880	6250		CLEO-3 Si
2001	1250		sc IR quads, e^+ target	

Table 2.   CLEO collaboration membership.
NSF and <u>DOE</u> supported.

Year →	1980	1990	2000
Cornell	+ − − − − − − − −−+ − − − − − − − − −−+ − −
<u>Harvard</u>	+ − − − − − − − −−+ − − − − − − − − −−+ − −
Ithaca	+ − − − − − − − −−+ − − − − − − − − −−+ − −
LeMoyne	+−
Rochester	+ − − − − − − − −+ − − − − − − − − −−+ − −
<u>Rutgers</u>	+ − − − −
Syracuse	+ − − − − − − − −−+ − − − − − − − −−+ − −
Vanderbilt	+ − − − − − − − −−+ − − − − − − − − −−+ − −
<u>Ohio State</u>	− − − − − − − −+ − − − − − − − − −−+ − −
Albany	− − − − − −+ − − − − − − − − −−+ − −
<u>Carnegie Mellon</u>	− − − − − − − − − −+ − −
Florida	− − − −− + − − − − − − − − −−+ − −
<u>Purdue</u>	− − − −+ − − − − − − − − − −−+ − −
<u>Minnesota</u>	− −− + − − − − − − − − −−+ − −
<u>Maryland</u>	− −
Kansas	− −+ − − − − − − − − −−+ − −
<u>Oklahoma</u>	− −+ − − − − − − − − −−+ − −
UC Santa Barbara	− + − − − − − − − − −−+ − −
<u>Colorado</u>	− + − − − − − − − −−
<u>Cal Tech</u>	− + − − − − − − − − −−+ − −
UC San Diego	− − − − − − − − −−+ − −
<u>So. Methodist</u>	− − − − − − − −−+ − −
McGill	− − − − − −−
Carleton	− − − − − − −−+ − −
<u>Illinois</u>	− − − − − − − + − −
Virginia Tech	− − − − − − − +−
<u>Hawaii</u>	− − − −−
<u>SLAC</u>	− − − −−
Wayne State	− − − −− + − −
UT Panamerican	+ − −
<u>UT Austin</u>	+ − −
Pittsburgh	+ − −
<u>Northwestern</u>	− −

Table 3.   CLEO officers.

	Spokesman	Analysis Coord.	Run Manager	Software Coord.
1979	A. Silverman, Cor	K. Berkelman, Cor	B. Gittelman, Cor	
1980	N. Horwitz, Syr	B. Gittelman, Cor E. Thorndike, Roc	E. Nordberg, Cor	
1981	E. Thorndike, Roc	M. Gilchriese, Cor	E. Nordberg, Cor	
1982	E. Thorndike, Roc	B. Gittelman, Cor	T. Ferguson, Cor	
1983	E. Thorndike, Roc	K. Berkelman, Cor	T. Ferguson, Cor	
1984	K. Berkelman, Cor	R. Kass, OSU	R. Galik, Cor	
1985	A. Silverman, Cor	M. Gilchriese, Cor	S. Gray, Cor	
1986	G. C. Moneti, Syr	T. Ferguson, CMU	S. Gray, Cor	D. Kreinick, Cor
1987	R. Galik, Cor	A. Jawahery, Syr	J. Kandaswamy, Cor	R. Namjoshi, Cor
1988	D. Cassel, Cor	S. Stone, Cor	B. Gittelman, Cor	R. Namjoshi, Cor
1989	R. Kass, OSU	Y. Kubota, Min	B. Gittelman, Cor	B. Heltsley, Cor
1990	E. Thorndike, Roc	D. Besson, Cor	J. Kandaswamy, Cor	B. Heltsley, Cor
1991	E. Thorndike, Roc	D. Besson, Cor	J. Kandaswamy, Cor	A. Weinstein, CIT
1992	D. Miller, Pur	D. Besson, Cor	R. Ehrlich, Cor	D. Kreinick, Cor
1993	D. Miller, Pur	T. Browder, Cor	R. Ehrlich, Cor	S. Patton, Min
1994	D. Miller, Pur	S. Menary, SBa	D. Cinabro, Har	D. Kreinick, Cor
1995	R. Poling, Min	R. Kutschke, SBa	M. Sivertz, SDi	S. Patton, Min
1996	R. Poling, Min	L. Gibbons, Roc	W. Ross, Okl	K. Lingel, SLAC
1997	G. Brandenburg, Har E. Thorndike, Roc	R. Briere, Har	M. Palmer, Ill	J. O'Neill, Min
1998	G. Brandenburg, Har E. Thorndike, Roc	F. Wuerthwein, CIT	B. Behrens, Col	R. Baker, Cor
1999	D. Cinabro, WSU K. Honscheid, OSU	D. Jaffe, SBa	G. Viehhauser, Syr	R. Baker, Cor
2000	J. Alexander, Cor J. Thaler, Ill	K. Ecklund, Cor	T. Pedlar, Ill	D. Kreinick, Cor
2001	J. Alexander, Cor I. Shipsey, Pur	K. Ecklund, Cor	D. Hennessy, Roc	J. Duboscq, Cor

Table 4.   CLEO refereed publications.
not including reviews, etc.
by year of submission.

Year, 19...	PRL	PRD	PLB	NIM..	TOTAL	authors
80	4				4	73
81	1				1	69
82	4	2	1	2	9	70
83	7	1	1		9	75
84	4	4			8	75
85	5	4		2	11	76
86	6	5	1		12	87
87	3	3	1		7	88
88	4	2			6	91
89	4	6	3		13	90
90	5	4	2		11	105
91	4	8			12	120
92	7	3	3	1	14	166
93	13	6	4		23	184
94	9	6	7	1	23	198
95	9	7	4		20	200
96	9	8	4		21	212
97	17	15	2	1	35	211
98	11	7	1		19	205
99	9	8			17	203
00	12	14	1		27	190

The following tables list the publications of the CLEO collaboration with submission dates up through the end of 2000. They are organized by subject matter, with each topic briefly identified. The journal names are abbreviated as follows: PRD and PRE are Physical Review D and E, PRL is Physical Review Letters, PLB is Physics Letters B, and NIM is Nuclear Instruments and Methods. The CLEO practice is to list authors alphabetically within an institution and to list institutions alphabetically but with the first listed institution rotating with each publication. CLEO papers therefore have a wide variety of first authors, and there is no real significance in the ordering of author names.

The papers can be accessed on the World Wide Web. The URL's are

    for PRD, http://prd.aps.org/

    for PRL, http://prl.aps.org/

    for PLB, http://www.elsevier.nl:80/inca/publications/store/

The PLB site is not accessible to non-subscribers. Prepublication versions of CLEO papers can be found on the web at

    http://www.lns.cornell.edu/public/CLEO/

    http://xxx.arXiv.cornell.edu/archive/hep-ex/

Table 5.   CLEO publications: Upsilons.

Topic	Author *et al.*	Reference	Submitted
CLEO-1			
$e^+e^- \to \Upsilon(1S,2S,3S)$	D. Andrews	PRL **44**, 1108(80)	15 Feb 80
$e^+e^- \to \Upsilon(4S)$	D. Andrews	PRL **45**, 219(80)	18 Apr 80
$\Upsilon(2S) \to \Upsilon(1S)\pi^+\pi^-$	J. Mueller	PRL **46**, 1181(81)	23 Feb 81
$\Upsilon(3S) \to \Upsilon(1S)\pi^+\pi^-$	J. Green	PRL **49**, 617(82)	16 Jun 82
$\Upsilon(1S,2S,3S) \to \ell^+\ell^-$	D. Andrews	PRL **50**, 807(83)	6 Jan 83
$\Upsilon(1S) \to \tau^+\tau^-$	R. Giles	PRL **50**, 877(83)	19 Jan 83
$e^+e^- \to X$	R. Giles	PRD **29**, 1285(84)	Nov 83
$\Upsilon(2S) \to \gamma X$	P. Haas	PRL **52**, 799(84)	21 Nov 83
$\Upsilon(2S) \to \Upsilon(1S)\pi^+\pi^-$	D. Besson	PRD **30**, 1433(84)	5 Mar 84
$\Upsilon(2S) \to \ell^+\ell^-$	P. Haas	PRD **30**, 1996(84)	20 Jul 84
$e^+e^- \to \Upsilon(5S,6S)$	D. Besson	PRL **54**, 381(85)	11 Oct 84
$\Upsilon(1S) \to gg\gamma$	S. Csorna	PRL **56**, 1222(86)	23 Dec 85
$\Upsilon(1S) \to \gamma X_{excl}$	A. Bean	PRD **34**, 905(86)	16 Jan 86
$\Upsilon(1S) \to \gamma X_{low\tau}$?	T. Bowcock	PRL **56**, 2676(86)	28 Feb 86
$\Upsilon(3S) \to \pi^+\pi^- X$	T. Bowcock	PRL **58**, 307(87)	29 Sep 86
CLEO-1.5			
$\Upsilon(1S,3S) \to \mu^+\mu^-$	W. Chen	PRD **39**, 3528(89)	14 Feb 89
$\Upsilon(1S) \to \psi X$	R. Fulton	PLB **224**, 445(89)	28 Apr 89
$\Upsilon(1S) \to \gamma X$	R. Fulton	PRD **41**, 1401(90)	5 May 89
$\Upsilon(4S) \to \psi X$ (non-$B\bar{B}$)	J. Alexander	PRL **64**, 2226(90)	17 Jan 90
$\Upsilon(3S) \to \pi^+\pi^- X$	I. Brock	PRD **43**, 1448(91)	24 Sep 90
CLEO-2			
$\Upsilon(3S) \to \chi_b(2P)X$	R. Morrison	PRL **67**, 1696(91)	30 May 91
$\Upsilon(3S) \to \chi_b(2P)X_{excl}$	G. Crawford	PLB **294**, 139(92)	26 Jun 92
$\Upsilon(3S)$ hadronic transitions	F. Butler	PRD **49**, 40(94)	8 Jul 93
$\Upsilon(1S) \to \tau^+\tau^-$	D. Cinabro	PLB **340**, 129 (94)	22 Sep 94
$\Upsilon(1S) \to \gamma X$	B. Nemati	PRD **55**, 5273(97)	30 Oct 96
$\Upsilon \to gg\gamma$ vs. $e^+e^- \to q\bar{q}\gamma$	M. S. Alam	PRD **56**, 17(97)	30 Dec 96
$\Upsilon(2S) \to \Upsilon(1S)h..$	J. Alexander	PRD **58**, 052004(98)	26 Feb 98
$\Delta m(\chi_{b,J})$	K. Edwards	PRD **59**, 032003(99)	12 Mar 98
$\Upsilon(1S) \to \gamma\pi\pi$	A. Anastassov	PRL **82**, 286(99)	5 Aug 98
$\Upsilon' \to \Upsilon\pi\pi$	S. Glenn	PRD **59**, 052003(99)	10 Aug 98
CLEO-2.5			
$\Upsilon(4S) \to B^+B^-$ vs. $B^0\bar{B}^0$	J. Alexander	PRL **86**, 2737(00)	1 Jun 00

Table 6.   CLEO publications: Soft hadronic physics and new particle searches.

Topic	Author *et al.*	Reference	Submitted
CLEO-1			
$\Upsilon(1S) \rightarrow$ axions?	M. S. Alam	PRD **27**, 1665(83)	22 Nov 82
$e^+e^- \rightarrow \xi(2200)X$?	S. Behrends	PLB **137**, 277(84)	29 Nov 83
$e^+e^- \rightarrow \Lambda X$	M. S. Alam	PRL **53**, 24(84)	12 Apr 84
$e^+e^- \rightarrow \pi X, KX, \ldots$	S. Behrends	PRD **31**, 2161(85)	18 Oct 84
Bose-Einstein correlations	P. Avery	PRD **32**, 2295(85)	11 Apr 85
$\Upsilon(1S) \rightarrow \zeta\gamma$?	D. Besson	PRD **33**, 300(86)	14 Aug 85
Magnetic monopoles?	T. Gentile	PRD **35**, 1081(87)	27 Oct 86
CLEO-1.5			
$B \rightarrow H^0 X$?	M. S. Alam	PRD **40**, 712(89)	13 Feb 89
Fractional charges?	T. Bowcock	PRD **40**, 263(89)	29 Mar 89
$\gamma\gamma \rightarrow X_{c\bar{c}}$	W. Chen	PLB **243**, 169(90)	27 Mar 90
CLEO-2			
gg and $q\bar{q} \rightarrow$ jets	M. S. Alam	PRD **46**, 4822(92)	1 Jun 92
$\gamma\gamma \rightarrow p\bar{p}$	M. Artuso	PRD **50**, 5484(94)	1 Sep 93
$\gamma\gamma \rightarrow \chi_{c2}$	J. Dominick	PRD **50**, 4265(94)	4 Oct 93
$\gamma\gamma \rightarrow \pi^+\pi^-$ or K^+K^-	J. Dominick	PRD **50**, 3027(94)	11 Mar 94
$\Upsilon(1S) \rightarrow \gamma$ neutralino?	R. Balest	PRD **51**, 2053(95)	11 Aug 94
$\gamma\gamma \rightarrow \Lambda\bar{\Lambda}$	S. Anderson	PRD **56**, R2485(97)	17 Jan 97
$\gamma\gamma \rightarrow f_J(2200)$?	R. Godang	PRL **79**, 3829(97)	18 Mar 97
$\eta \rightarrow e^+e^-$?	T. Browder	PRD **56**, 5359(97)	3 Jun 97
$\sigma_{tot}(e^+e^- \rightarrow h..)$ at 10.52 GeV	R. Ammar	PRD **57**, 1350(98)	7 Jul 97
$F_{PS,\gamma}$ at high q^2	J. Gronberg	PRD **57**, 33(98)	12 Jul 97
CLEO-2.5			
$\gamma\gamma \rightarrow f_J(2200)$?	M. S. Alam	PRL **81**, 3328(98)	28 May 98
$\eta' \rightarrow$ rare?	R. A. Briere	PRL **84**, 26(00)	22 Jul 99
$\eta_c: m, \Gamma, \Gamma_{\gamma\gamma}$	G. Brandenburg	PRL **85**, 3095(00)	20 Jun 00
$e^+e^- \rightarrow \bar{\bar{b}}\bar{b}$?	V. Savinov	PRD **63**, R051101(01)	17 Oct 00

Table 7. CLEO publications: Tau lepton hadronic decay modes.

Topic	Author *et al.*	Reference	Submitted
CLEO-1			
$\tau \to \eta X$ and ωX	P. Baringer	PRL **59**, 1993(87)	29 Jul 87
CLEO-1.5			
$\tau \to K^* X$	M. Goldberg	PLB **251**, 223(90)	30 Jul 90
CLEO-2			
$\tau \to \eta X$	M. Artuso	PRL **69**, 3278(92)	4 Aug 92
$\tau \to h\pi^0$'s	M. Procario	PRL **70**, 1207(93)	14 Oct 92
$\tau \to \pi\pi\pi\pi^0\pi^0\nu$	D. Bortoletto	PRL **71**, 1791(93)	6 Jul 93
Cabibbo suppressed decay	M. Battle	PRL **73**, 1079(94)	4 Feb 94
$\tau \to h\pi^0\nu$	M. Artuso	PRL **72**, 3762(94)	1 Apr 94
$\tau \to 5\pi$	D. Gibaut	PRL **73**, 934(94)	22 Apr 94
$\tau \to 3h^\pm\nu, 3h^\pm\pi^0\nu$	R. Balest	PRL **75**, 3809(95)	14 Jul 95
$\tau \to K_S^0..$	T. E. Coan	PRD **53**, 6037(96)	10 Jan 96
$\tau \to K\eta\nu$	J. Bartelt	PRL **76**, 4119(96)	12 Jan 96
$\tau \to \phi X$	P. Avery	PRD **55**, R1119(97)	15 Oct 96
$\tau \to 3\pi\eta\nu, f_1\pi\nu$	T. Bergfeld	PRL **79**, 2406(97)	25 Jun 97
$\tau \to 7\pi^\pm\pi^0\nu$?	K. Edwards	PRD **56**, R5297(97)	9 Jul 97
$\tau \to 5\pi^\pm\pi^0\nu$	S. Anderson	PRL **79**, 3814(97)	9 Jul 97
$\tau \to K^{*-}\eta\nu$	M. Bishai	PRL **82**, 281(99)	15 Sep 98
$\tau \to$ 3-prong with $K^\pm$	S. Richichi	PRD **60**, 112002(99)	15 Oct 98
$\tau \to \pi^\pm 2\pi^0\nu$, ν-helicity	D. M. Asner	PRD **61**, 012002(00)	16 Feb 99
CLEO-2.5			
$\tau \to \pi^\pm 2\pi^0\nu$ h-structure	T. Browder	PRD **61**, 052004(00)	16 Aug 99
$\tau \to 3\pi^\pm\pi^0\nu$ resonances	K. Edwards	PRD **61**, 072003(00)	8 Sep 99
$\tau \to \pi^\pm\pi^0\nu$ h-structure	S. Anderson	PRD **61**, 112002(00)	21 Oct 99
$\tau \to K^\pm 2\pi^\pm\nu$ resonances	D. M. Asner	PRD **6s**, 072006(00)	25 Apr 00

Table 8.   CLEO publications: Other tau lepton papers.

Topic	Author *et al.*	Reference	Submitted
CLEO-1			
Michel parameter	S. Behrends	PRD **32**, 2468(85)	15 Jul 85
$M(\nu_\tau)$	S. Csorna	PRD **35**, 2747(87)	3 Nov 86
$\tau(\tau)$	C. Bebek	PRD **36**, 690(87)	13 Apr 87
	P. Baringer	PRL **59**, 1993(87)	29 Jul 87
CLEO-1.5			
$\tau \rightarrow$ no-ν?	T. Bowcock	PRD **41**, 805(90)	13 Oct 89
$\tau \rightarrow e\nu\bar{\nu}$	R. Ammar	PRD **45**, 3976(92)	2 Dec 91
CLEO-2			
$\tau(\tau)$	M. Battle	PLB **291**, 488(92)	22 Jul 92
$\tau \rightarrow \gamma\mu$?	A. Bean	PRL **70**, 138(93)	24 Sep 92
$\tau \rightarrow e\nu\bar{\nu}$	D. Akerib	PRL **69**, 3610(92)	28 Sep 92
$M(\tau)$	R. Balest	PRD **47**, 3671(93)	9 Feb 93
$\tau \rightarrow$ no-ν?	J. Bartelt	PRL **73**, 1890(94)	6 Jun 94
α_S from τ decays	T. Coan	PLB **365**, 580(95)	19 Jun 95
$\tau \rightarrow 3\ell2\nu$?	M. S. Alam	PRL **76**, 2637(96)	22 Nov 95
$\tau(\tau)$	R. Balest	PLB **388**, 402(96)	6 Jul 96
ℓ universality	A. Anastassov	PRD **55**, 2559(97)	6 Nov 96
$\tau \rightarrow e\gamma, \mu\gamma$?	K. Edwards	PRD **55**, R3919(97)	12 Nov 96
Michel parameters	R. Ammar	PRL **78**, 4686(97)	26 Dec 96
ν-helicity from $E(h)$ correl.	T. E. Coan	PRD **55**, 7291(97)	22 Jan 97
$\tau \rightarrow \pi^0, \eta,$ no-ν?	G. Bonvicini	PRL **79**, 1221(97)	17 Apr 97
Michel parameters, ν-helicity	J. Alexander	PRD **56**, 5320(97)	15 May 97
$\tau \rightarrow$ no-ν?	B. Nemati	PRD **57**, 5903(98)	8 Dec 97
$M(\nu_\tau)$	R. Ammar	PLB **431**, 209(98)	3 Apr 98
CP in τ decay	S. Anderson	PRL **81**, 3823(98)	21 May 98
$\tau \rightarrow$ B or L violating	R. Godang	PRD **59**, 091303(99)	15 Dec 98
CLEO-2.5			
$M(\nu_\tau)$ from $\tau \rightarrow 3\pi^\pm\nu$	M. Athenas	PRD **61**, 052002(00)	4 Jun 99
$\tau \rightarrow \ell\gamma\nu$	T. Bergfeld	PRL **84**, 830(00)	7 Sep 99
$\tau \rightarrow \mu\gamma$?	S. Ahmed	PRD **61**, 071101(00)	25 Oct 99

Table 9.   CLEO publications: D meson hadronic decay modes.

Topic	Author *et al.*	Reference	Submitted
CLEO-1			
$D^0 \to \overline{K^0}\phi$	C. Bebek	PRL **56**, 1893(86)	5 Feb 86
CLEO-1.5			
$D^- \to K\bar{K}$ or $\pi\bar{\pi}$	J. Alexander	PRL **65**, 1184(90)	13 Jun 90
$D^0 \to \pi^0 X$ or ηX	K. Kinoshita	PRD **43**, 2836(91)	18 Dec 90
CLEO-2			
$D^0 \to \overline{K^0}$ and $\overline{K^{*0}}$	M. Procario	PRD **48**, 4007(93)	14 Oct 92
$D \to \pi\pi$	M. Selen	PRL **71**, 1973(93)	11 Jun 93
$D^0 \to K^-\pi^+$	D. Akerib	PRL **71**, 3070(93)	23 Aug 93
$D^0 \to K^+\pi^-$	D. Cinabro	PRL **72**, 1406(94)	2 Dec 93
$D^+ \to K^-\pi^+\pi^+$	R. Balest	PRL **72**, 2328(94)	17 Jan 94
D^0 FCNC decays	A. Freyberger	PRL **76**, 3065(96)	10 Jan 96
$D^0 \to K^-\pi^+\pi^0$	B. Barish	PLB **373**, 335(96)	5 Feb 96
$D^0 \to K\overline{K}X$	D. M. Asner	PRD **54**, 4211(96)	16 Apr 96
$D^+ \to K^0_S K^+$, $K^0_S \pi^+$	M. Bishai	PRL **78**, 3261(97)	27 Dec 96
$D^0 \to K^-\pi^+$ via partial D^{*+}	M. Artuso	PRL **80**, 3193(98)	17 Dec 97
CLEO-2/5			
$D^0 - \bar{D}^0$ mixing?	R. Godang	PRL **84**, 5038(00)	3 Jan 00
$D^0 \to K^-\pi^+\pi^0$ Dalitz	S. Kopp	PRD **63**, 092001(01)	17 Nov 00
CP in $D^0 \to K_S/\pi^0 K_S/\pi^0$?	G. Bonvicini	PRD **63**, 0701101(01)	19 Dec 00

Table 10. CLEO publications: Other D meson papers.

Topic	Author *et al.*	Reference	Submitted
CLEO-1			
$e^+e^- \to D^+X$	C. Bebek	PRL **49**, 610(82)	26 May 82
$e^+e^- \to D^{(*)}X$	P. Avery	PRL **51**, 1139(83)	21 Jul 83
$\tau(D^0, D^+, D_s)$	S. Csorna	PLB **191**, 318(87)	29 Jan 87
$e^+e^- \to c\bar{c}$	T. Bowcock	PRD **38**, 2679(88)	26 Feb 88
$e^+e^- \to c\bar{c}$	D. Bortoletto	PRD **37**, 1719(88)	26 Oct 88
$D \to \ell^+\ell^-X$?	P. Haas	PRL **60**, 1614(88)	23 Nov 88
CLEO-1.5			
$e^+e^- \to D_J X$	P. Avery	PRD **41**, 774(90)	24 Aug 89
$e^+e^- \to \vec{D}^{*+}X$	Y. Kubota	PRD **44**, 593(91)	25 Jan 91
$D \to$ "unusual"	R. Ammar	PRD **44**, 3383(92)	22 Apr 91
$D^0 \to K^{*-}e\nu$ and $K^-e\nu$	G. Crawford	PRD **44**, 3394(92)	10 May 91
CLEO-2			
$D^* \to D\pi$ and $D\gamma$	F. Butler	PRL **69**, 2041(92)	15 Jun 92
$M(D^*) - M(D)$	D. Bortoletto	PRL **69**, 2046(92)	13 Jul 92
$D^+ \to \pi^0 \ell\nu$	M. S. Alam	PRL **71**, 1311(93)	16 Jun 93
$D \to X_{excl}\ell\nu$	A. Bean	PLB **317**, 647(93)	30 Sep 93
$e^+e^- \to D_1^0 X$ and $D_2^{*0}X$	P. Avery	PLB **331**, 236(94)	8 Mar 94
$e^+e^- \to D_1^+ X$ and $D_2^{*+}X$	T. Bergfeld	PLB **340**, 194(94)	30 Sep 94
$D^0 \to \pi^- e^+\nu$	F. Butler	PRD **52**, 2656(95)	27 Jan 95
CP in D^0 decays	J. Bartelt	PRD **52**, 4860(95)	22 May 95
$D^0 \to Xe\nu$	Y. Kubota	PRD **54**, 2994(96)	25 Oct 95
$D^+ \to \pi^0 \ell^+\nu, \eta e^+\nu$	J. Bartelt	PLB **405**, 373(97)	1 Apr 97
$D^{*+} \to D^+\gamma$	J. Bartelt	PRL **80**, 3919(98)	19 Nov 97
$D^{*\pm}$ spin alignment	G. Brandenburg	PRD **58**, 052003(98)	26 Feb 98
CLEO-2.5			
$\tau(D_{(s)})$	G. Bonvicini	PRL **82**, 4586(99)	8 Feb 99

Table 11.   CLEO publications: D_s charmed mesons.

Topic	Author *et al.*	Reference	Submitted
CLEO-1			
$e^+e^- \to D_s X$	A. Chen	PRL **51**, 634(83)	22 Jun 83
CLEO-1.5			
$D_s \to X_{excl}$	W. Chen	PLB **226**, 192 (89)	19 May 89
$D_s \to \phi\ell\nu$	J. Alexander	PRL **65**, 1531(90)	28 Jun 90
CLEO-2			
$D_s \to \eta^{(\prime)}\pi$	J. Alexander	PRL **68**, 1275(92)	27 Sep 91
$D_s \to \eta^{(\prime)}\rho$	P. Avery	PRL **68**, 1279(92)	27 Sep 91
$D_s \to \eta^{(\prime)}\pi$ and $\eta^{(\prime)}\rho$	M. Daoudi	PRD **45**, 3965(92)	30 Sep 91
$e^+e^- \to D_{s1}^+ X$	J. Alexander	PLB **303**, 378(93)	5 Feb 93
$D_s \to \mu\nu$	D. Acosta	PRD **49**, 5690(94)	3 Aug 93
$e^+e^- \to D_{s2}^{*+} X$	Y. Kubota	PRL **72**, 1972(94)	14 Jan 94
$M(D_s^{*+}) - M(D_s^+)$	D. Brown	PRD **50**, 1884(94)	27 Jan 94
$D_s \to \phi\ell\nu$	F. Butler	PLB **324**, 255(94)	1 Feb 94
$D_s \to \phi e\nu$ form factors	P. Avery	PLB **337**, 405(94)	27 Jul 94
$D_s^{*+} \to D_s^+ \pi^0$	J. Gronberg	PRL **75**, 3232(95)	21 Jul 95
$D_s^+ \to \eta^{(\prime)}\ell^+\nu$	G. Brandenburg	PRL **75**, 3804(95)	24 Jul 95
$D_s \to \phi\pi^\pm$	M. Artuso	PLB **378**, 364(96)	2 Feb 96
$D_s \to \omega\pi^\pm$	R. Balest	PRL **79**, 1436(97)	1 May 97
$D_s \to \mu\nu$ for f_{D_s}	M. Chada	PRD **58**, 032002(98)	10 Dec 97
$D_s \to \eta^{(\prime)}\pi^\pm, \eta^{(\prime)}\rho^\pm$	C. P. Jessop	PRD **58**, 052002(98)	31 Dec 97
CLEO-2.5			
$c \to D_s^{(*)}$	R. Briere	PRD **62**, 072003(00)	25 Apr 00

Table 12. CLEO publications: Charmed baryons.

Topic	Author *et al.*	Reference	Submitted
CLEO-1			
$e^+e^- \to \Lambda_c X$	T. Bowcock	PRL **55**, 923(85)	5 Jun 85
$e^+e^- \to \Xi_c^0 X$	P. Avery	PRL **62**, 863(89)	21 Nov 88
$e^+e^- \to \Sigma_c^{++,0} X$	T. Bowcock	PRD **40**, 1240(89)	13 Dec 88
CLEO-1.5			
Λ_c decay asymmetry	P. Avery	PRL **65**, 2842(90)	10 Aug 90
$e^+e^- \to \Lambda_c X$	R. Fulton	PRD **43**, 3599(91)	27 Aub 90
CLEO-2			
$\Xi_c \to \Omega K$	S. Henderson	PLB **283**, 161(92)	15 Jan 92
$\Lambda_c \to \Xi K, \Sigma K K, \Xi K \pi$	P. Avery	PRL **71**, 2391(93)	6 May 93
$\Lambda_c \to \Sigma^+ \pi, \omega$, etc.	Y. Kubota	PRL **71**, 3255(93)	24 Jun 93
$e^+e^- \to \Sigma_c^+ X$	G. Crawford	PRL **71**, 3259(93)	24 Jun 93
$\Lambda_c \to \Lambda \pi\pi, \Sigma^0 n\pi$	P. Avery	PLB **325**, 257(94)	17 Dec 93
$\Lambda_c \to \Lambda \ell \nu$	T. Bergfeld	PLB **323**, 219(94)	20 Jan 94
$\Xi_c \to \Xi e \nu$	J. Alexander	PRL **74**, 3113(95)	12 Oct 94
$\Lambda_c \to \eta X_{excl}$ etc.	R. Ammar	PRL **74**, 3534(95)	10 Nov 94
$\Lambda_c^*(2593, 2625) \to \Lambda_c \pi^+ \pi^-$	K. Edwards	PRL **74**, 3331(95)	21 Nov 94
$\Lambda_c \to \Lambda e \nu$	G. Crawford	PRL **75**, 624(95)	13 Jan 95
$\Lambda_c \to \Lambda \pi^\pm, \Sigma'^+ \pi^0$ asyms.	M. Bishai	PLB **350**, 256(95)	22 Feb 95
$\Lambda_c \to p\phi$	J. Alexander	PRD **53**, 1013(96)	25 Jul 95
$\Xi_c^* \to \Xi_c^+ \pi^-$	P. Avery	PRL **75**, 4364(95)	15 Aug 95
$\Xi_c^+ \to \Sigma^+ K^- \pi^+, \Lambda K^- \pi^+ \pi^-$	T. Bergfeld	PLB **365**, 431(96)	7 Nov 95
$\Xi_c^+ \to$ new modes	K. Edwards	PLB **373**, 261(96)	23 Jan 96
$\Xi_c^* \to \Xi_c^0 \pi^+$	L. Gibbons	PRL **77**, 810(96)	1 Mar 96
$\Sigma_c^* \to \Lambda_c \pi^\pm$	G. Brandenburg	PRL **78**, 2304(97)	26 Sep 96
$\Lambda_c \to p\overline{K}\pi..$	M. S. Alam	PRD **57**, 4467(98)	10 Sep 97
$\Xi_c^* \to \Xi_c^{+,0} \gamma$	C. P. Jessop	PRL **82**, 492(99)	19 Oct 98
CLEO-2.5			
$\Xi_c' \to \Xi_c^* \pi$	J. Alexander	PRL **83**, 3390(99)	8 Jun 99
$\Lambda_c \to p K^- \pi^+$	D. E. Jaffe	PRD **62**, 072005(00)	28 Mar 00
$c \to \Theta_c \to \Lambda X$	R. Ammar	PRD **62**, 092007(00)	28 Apr 00
$\Sigma_c^{*+}, M(\Sigma_c^+)$	R. Ammar	PRL **86**, 1167(01)	19 Jul 00
Ω_c^0 observation	D. C.-Hennessy	PRL **86**, 3730(01)	11 Oct 00
$\tau(\Lambda_c)$	A. Mahmood	PRL **86**, 2232(01)	15 Nov 00

Table 13.   CLEO publications: B leptonic decays to charm.

Topic	Author *et al.*	Reference	Submitted
CLEO-1			
$b \to \ell\ell q$, etc.?	A. Chen	PLB **122**, 317(83)	20 Dec 92
$B \to \ell^+\ell^- X$	P. Avery	PRL **53**, 1309(84)	25 May 84
$\tau(B^0)/\tau(B^+)$, $B^0 \rightleftharpoons \overline{B^0}$?	A. Bean	PRL **58**, 183(87)	24 Jul 86
$B \to \ell^+\ell^- X$	A. Bean	PRD **35**, 3533(87)	1 Apr 87
CLEO-1.5			
$B^0 \rightleftharpoons \overline{B^0}$	M. Artuso	PRL **62**, 2233(89)	16 Feb 89
$\bar{B} \to D^*\ell\nu$	D. Bortoletto	PRL **63**, 1667(89)	14 Jun 89
$\bar{B} \to D\ell\nu$, $DX\ell\nu$	R. Fulton	PRD **43**, 651(91)	1 Aug 90
$\bar{B}^{0,-} \to X\ell\nu$	S. Henderson	PRD **45**, 2212(92)	8 Jul 91
CLEO-2			
$\theta(\ell)$ in $\bar{B} \to X\ell\nu$	S. Sanghera	PRD **47**, 791(93)	28 Jul 92
$B^0 \rightleftharpoons \overline{B^0}$	J. Bartelt	PRL **71**, 1680(93)	29 Apr 93
$\bar{B} \to D^*\ell\nu$	B. Barish	PRD **51**, 1014(95)	23 Jun 94
$\bar{B}^{0,-} \to X\ell\nu$	M. Athenas	PRL **73**, 3503(94)	24 Jun 94
$B \to X\ell\nu$ with ℓ-tag	B. Barish	PRL **76**, 1570(96)	16 Oct 95
$B^0 \to D^{*+}\ell^-\bar{\nu}$	J. Duboscq	PRL **76**, 3898(96)	27 Nov 95
$B \to X\ell\nu$ spectrum	M. Artuso	PLB **399**, 321(97)	17 Feb 97
$B \to D\ell\nu$ Γ, F	M. Athenas	PRL **79**, 2208(97)	30 May 97
$B \to D^{**}\ell\nu$	A. Anastasssov	PRL **80**, 4127(98)	18 Aug 97
$B \to \Theta_c\ell\nu$	G. Bonvicini	PRD **57**, 6604(98)	2 Dec 97
$B \to D\ell\nu$ $\mathcal{B}r$, F	J. Bartelt	PRL **82**, 3020(99)	25 Nov 98
CLEO-2.5			
$B^0 - \bar{B}^0$ mixing parameters	B. Behrens	PLB **490**, 36(00)	10 Aug 00

Table 14. CLEO publications: B charmless leptonic decays.

Topic	Author *et al.*	Reference	Submitted
CLEO-1			
$B \to e\nu$	C. Bebek	PRL **46**, 84(81)	1 Oct 80
$B \to \mu\nu$	K. Chadwick	PRL **46**, 88(81)	31 Oct 80
$B \to \ell\nu$	K. Chadwick	PRD **27**, 475(83)	9 Aug 82
$b \to u\ell\nu$?	A. Chen	PRL **52**, 1084(84)	30 Jan 84
$b \to u\ell\nu$?	S. Behrends	PRL **59**, 407(87)	4 May 87
CLEO-1.5			
$b \to u\ell\nu$	R. Fulton	PRL **64**, 16(90)	8 Nov 89
CLEO-2			
$B \to \rho\ell\nu, \pi\ell\nu$?	A. Bean	PRL **70**, 2681(93)	21 Dec 92
$b \to u\ell\nu$	J. Bartelt	PRL **71**, 4111(93)	7 Sep 93
$\bar{B} \to \ell^+\ell^-$?	R. Ammar	PRD **49**, 5701(94)	1 Dec 93
$B \to \ell\nu$?	M. Artuso	PRL **75**, 785(95)	31 Mar 95
$B \to \pi\ell\nu, \rho\ell\nu$	J. Alexander	PRL **77**, 5000(96)	1 Jul 96
$B \to \ell\nu\gamma$?	T. Browder	PRD **56**, 11(97)	12 Nov 96
$b \to s\ell^+\ell^-$?	S. Glenn	PRL **80**, 2289(98)	1 Oct 97
CLEO 2.5			
$B \to \rho\ell\nu$ for $|V_{ub}|$	B. Behrens	PRD **61**, 052001(00)	24 May 99
$B \to \ell^+\ell^-$?	T. Bergfeld	PRD **62**, R091102(00)	19 Jul 00
$B \to \tau\nu, K\nu\bar{\nu}$	T. Browder	PRL **86**, 2950(01)	26 Jul 00

Table 15.   CLEO publications: B nonleptonic decays to charmonium and to baryons.

Topic	Author *et al.*	Reference	Submitted
CLEO-1			
$B \to$ baryons	M. S. Alam	PRL **51**, 1143(83)	24 Jun 83
$B \to \psi X$	P. Haas	PRL **55**, 2348(85)	26 Jun 85
$B \to \psi X$	M. S. Alam	PRD **34**, 3279(86)	2 Sep 86
$B \to Y_{c,baryon}$	M. S. Alam	PRL **59**, 22(87)	28 Apr 87
CLEO-1.5			
$B \to$ baryons	G. Crawford	PRD **45**, 752(92)	15 Apr 91
CLEO-2			
$\bar{B} \to \Sigma_c X$	M. Procario	PRL **73**, 1472(94)	21 Dec 93
$\bar{B} \to X_{c,excl}$ and $X_{c\bar{c},excl}$	M. S. Alam	PRD **50**, 43(94)	4 Jan 94
$B \to \psi\pi$	J. Alexander	PLB **341**, 435(95)	14 Oct 94
$B \to c\bar{c}X$	R. Balest	PRD **52**, 2661(95)	13 Dec 94
$B \to \psi\rho$	M. Bishai	PLB **369**, 186(96)	11 Dec 95
$B \to$ baryons	R. Ammar	PRD **55**, 13(97)	12 Jun 96
$B \to \psi K^{(*)}$	C. P. Jessop	PRL **79**, 4533(97)	24 Feb 97
$B \to \Xi_c^{0,+} X$	B. Barish	PRL **79**, 3599(97)	8 May 97
$B \to \Theta_c X$	X. Fu	PRL **79**, 3125(97)	12 Jul 97
$B \to$ baryons, rare	T. E. Coan	PRL **82**, 492(99)	19 Oct 98
$B \to \psi\phi K$	A. Anastassov	PRL **84**, 1393(00)	3 Nov 99
$M(B)$ from $B \to \psi^{(\prime)} K$	S. E. Csorna	PRD **61**, R111101(00)	5 Jan 00
CP in $B^{\pm} \to \psi^{(\prime)} K^{\pm}$	G. Bonvicini	PRL **84**, 5940(00)	2 Mar 00
$B \to (c\bar{c})..$	P. Avery	PRD **62**, R051101(00)	28 Apr 00
$B \to \eta_c K, \chi_{c0} K$	K. Edwards	PRL **86**, 30(01)	7 Jul 00
$B \to \psi(2S) K^{(*)}$	S. Richichi	PRD **63**, R031103(01)	16 Sep 00
$B \to \chi_{c1,2} X$	S. Chen	PRD **63**, R031102(01)	19 Sep 00

Table 16.   CLEO publications: Other B nonleptonic decays to charm.

Topic	Author *et al.*	Reference	Submitted
CLEO-1			
$B \to K^{\pm}X,\ K_S X$	A. Brody	PRL **48**, 1070(82)	28 Jan 82
$\langle n_{ch} \rangle$ in $B \to X$	M. S. Alam	PRL **49**, 357(82)	26 May 92
$B \to X_{excl}$	S. Behrends	PRL **50**, 881(83)	24 Jan 83
$\bar{B} \to D^0 X$	J. Green	PRL **51**, 347(83)	20 May 83
$B \to X_{excl}$	R. Giles	PRD **30**, 2279(84)	10 Sep 84
$\bar{B} \to D^{*+}X$	S. Csorna	PRL **54**, 1894(85)	11 Feb 85
$\bar{B} \to D^* \rho$	A. Chen	PRD **31**, 2386(85)	25 Feb 85
$B \to \phi X$	D. Bortoletto	PRL **56**, 800(86)	21 Oct 85
$\bar{B} \to D_s X$	P. Haas	PRL **56**, 2781(86)	7 Apr 86
$\bar{B} \to X_c$	D. Bortoletto	PRD **35**, 19(87)	21 Jul 86
$B \to K^{+,-,0}X$	M. S. Alam	PRL **58**, 1814(87)	29 Dec 86
$B \to X_{excl}$	C. Bebek	PRD **36**, 3533(87)	26 May 87
CLEO-1.5			
$\bar{B} \to D_s X$	D. Bortoletto	PRL **64**, 2117(90)	15 Jan 90
CLEO-2			
$e^+ e^- \to B^* X$	D. Akerib	PRL **67**, 1692(91)	20 May 91
$\bar{B} \to X_c$ and $X_{c\bar{c}}$	D. Bortoletto	PRD **45**, 21(92)	29 Jul 92
$B \to \eta X$	Y. Kubota	PLB **350**, 256(95)	22 Feb 95
$B \to D_s X$	D. Gibaut	PRD **53**, 4734(96)	19 Oct 95
$B \to 3h^{\pm}$	T. Bergfeld	PRL **77**, 4503(96)	23 Aug 96
$B \to D^{(*)}X$	L. Gibbons	PRD **56**, 3783(97)	28 Feb 97
$B^0 \to D^{*+}D^{*-}$?	D. M. Asner	PRL **79**, 799(97)	23 Apr 97
$B \to D^* \pi$	G. Brandenburg	PRL **80**, 2762(98)	25 Jun 97
$B \to$ color suppressed	M. Bishai	PRD **57**, 5363(98)	28 Aug 97
$B \to$ charm	T. E. Coan	PRL **80**, 1150(98)	12 Oct 97
$B \to D_{s1}(2536)X$	M. Bishai	PRD **57**, 3847(98)	23 Oct 97
$B^+ \to \bar{D}^0 K^+$	M. Athenas	PRL **80**, 5493(98)	3 Mar 98
$B^0 \to D^{*+}D^{*-}$	M. Artuso	PRL **82**, 3020(99)	16 Nov 98
CLEO-2.5			
$\langle n_{ch} \rangle$ in B decay	G. Brandenburg	PRD **61**, 072002(00)	8 Sep 99
$B \to \gamma..$	T. E. Coan	PRL **84**, 5283(00)	23 Dec 99
$\bar{B}^0 \to D^{*0}\gamma$?	M. Artuso	PRL **84**, 4392(00)	3 Jan 00
$B^0 \to D^{*+}D^{*-}$	E. Lipeles	PRD **62**, 032005(00)	29 Feb 00
$B \to D_s^{(*)}D^{*(*)}$	S. Ahmed	PRD **62**, 112003(00)	8 Aug 00
$B^0 \to D^{*-}p\bar{p}\pi^+,\ D^{*-}p\bar{n}$	S. Anderson	PRL **86**, 2732(01)	5 Sep 00

Table 17.   CLEO Publications: B charmless nonleptonic decays.

Topic	Author *et al.*	Reference	Submitted
CLEO-1			
$b \to u, s$ excl.?	P. Avery	PLB **183**, 429(87)	6 Nov 86
CLEO-1.5			
$B \to p\bar{p}\pi$? or $p\bar{p}\pi\pi$?	C. Bebek	PRL **62**, 8(89)	3 Oct 88
$b \to u$ excl.?	D. Bortoletto	PRL **62**, 2436(89)	16 Feb 89
$b \to s$ excl.?	P. Avery	PLB **223**, 470(89)	10 Mar 89
CLEO-2			
$B \to K^*\gamma$	R. Ammar	PRL **71**, 674(93)	24 May 93
$B \to \pi\pi, K\pi$	M. Battle	PRL **71**, 3922(93)	11 Aug 93
$b \to u$ with $D_s^{(*)}$?	J. Alexander	PLB **319**, 365(93)	8 Nov 93
$b \to s\gamma$	M. S. Alam	PRL **74**, 2885(95)	13 Dec 94
$B \to$ charmless exclusive	D. M. Asner	PRD **53**, 1039(96)	19 Jul 95
$B \to K\pi, \pi\pi, KK$	R. Godang	PRL **80**, 3456(98)	17 Nov 97
$B \to \eta^{(\prime)}$.. 2-body	B. Behrens	PRL **80**, 3710(98)	5 Jan 98
$B^+ \to \omega K^+$	T. Bergfeld	PRL **81**, 272(98)	20 Mar 98
$B \to \eta' X$	T. Browder	PRL **81**, 1786(98)	28 Apr 98
CLEO-2.5			
$B \to \eta^{(\prime)}$.. 2-body	S. Richichi	PRL **85**, 520(00)	23 Dec 99
CP in $B \to$ charmless	S. Chen	PRL **85**, 525(00)	23 Dec 99
$B \to K^{\pm,0}\pi^0, \pi^+\pi^-$	D. Cronin-H.	PRL **85**, 515(00)	27 Dec 99
$B \to$ charmless PV	C. P. Jessop	PRL **85**, 2881(00)	30 May 00
$B \to \phi K^{(*)}$	R. A. Briere	PRL **86**, 3718(01)	18 Jan 01

Table 18.   CLEO publications: Instrumentation.

Topic	Author *et al.*	Reference	Submitted
CLEO-1			
SC solenoid	D. Andrews	Adv. Cryo. Eng. **27**, 143(82)	
CLEO-1	D. Andrews	NIM **A211**, 47(83)	23 Aug 82
CLEO-1.5			
Drift chamber DR2	D. Cassel	NIM **A252**, 325(86)	
CLEO-2			
CsI calorimeter	E. Blucher	NIM **A249**, 201(86)	19 Mar 86
Trigger	C. Bebek	NIM **A302**, 261(91)	5 Dec 90
CLEO-2	Y. Kubota	NIM **A320**, 66(92)	16 Jan 92
Luminosity measurement	G. Crawford	NIM **A345**, 429(94)	1 Feb 94
CLEO-2.5			
Dynamic β effect	D. Cinabro	PRE **57**, 1193(98)	30 Jun 97

References

[1] M. S. Livingston and J. P. Blewett, *Particle Accelerators*, McGraw-Hill, 1962. This is probably the first text on particle accelerators, with plenty of detail on the earliest machines.

[2] R. R. Wilson, *An Anecdotal Account of Accelerators at Cornell (For Bethe's Festschrift)*, May 18, 1966 (unpublished). This is a good source for the history of the Laboratory of Nuclear Studies from 1945 to 1966.

[3] A. Silverman, *Birth of Electron Synchrotrons at Cornell*, CLNS 88/875 (unpublished). This covers the same period as Wilson's anecdotal account, but with more emphasis on the experimental data.

[4] B. D. McDaniel and A. Silverman, "The 10-GeV synchrotron at Cornell," *Phys. Today* **21** (Oct 1968) 29. This covers the design and construction of the synchrotron and the early experimental program.

[5] K. Gottfried, *Heavy Quark Spectroscopy Before the Discovery of Υ*, CLNS-97/1411, presented at the $b20$ Symposium, Illinois Institute of Technology, 29 Jun–2 Jul 97.

[6] *Design Report: Cornell Electron Storage Ring, April 1977*, CLNS 360.

[7] A. Abashian, "Recollections on the birth of CESR," presented at a talk at Cornell University, 9 Dec 99. This details the events at the National Science Foundation leading to the approval and funding for CESR construction.

[8] B. D. McDaniel, "The commissioning and performance characteristics of CESR," *IEEE Trans. Nucl. Sci.* **NS-28**, No. 3, (Jun 81). CESR in the first year of operations.

[9] D. M. Kaplan, *The Discovery of the Upsilon Family*, paper presented at the Int. Conf. the History of Original Ideas and Basic Discoveries in Particle Physics, Erice, Italy, 29 Jul–3 Aug 94. An account of the discovery of the $\Upsilon(1S)$ and $\Upsilon(2S)$ (and perhaps $\Upsilon(3S)$ at Fermilab).

[10] G. Salvini and A. Silverman, "Physics with matter–antimatter colliders," *Phys. Rep.*, **171** (1988) 231.

[11] J. Warnow-Blewett *et al.*, *AIP Study of Multi-Institutional Collaborations, Phase I: High-Energy Physics*, Center for History of Physics, American Institute of Physics, 1992. See especially the *Probe Report on the CLEO Experiment at CESR* (Report No. 4, Part D) by Joel Genuth.

[12] R. A. Burnstein, D. M. Kaplan and H. A. Rubin, (eds.) *Twenty Beautiful Years of Bottom Physics*, AIP Conference Proc. 424. Presentations at a symposium on the history of *b*-physics held at Illinois Institute of Technology in 1997. See in particular the talks by Silverman, Stone, Lee-Franzini, Poling and Honscheid.

[13] K. Berkelman, "Upgrading CESR," *Beam Line* (publ. by SLAC) **27**, No. 2, (summer 1997) 18. In addition to the above references cited in the text, there are the following unpublished sources of detailed historical information on CESR and CLEO.

[14] Quarterly and annual reports from CESR/CLEO to the NSF.

[15] Annual funding proposals from CESR/CLEO to the NSF.

[16] Presentations to the CESR Program Advisory Committee.

[17] Minutes of monthly CLEO collaboration meetings, N. Horwitz, Syracuse University.

[18] Internal CESR and CLEO reports. The important series are denoted CLNS, CBX, CBN, SRF.

[19] `http://www.lns.cornell.edu`   Current information on the Laboratory facilities, research program, publications, etc.

Index